Shades of Blue

Shades of Blue

The Uniforms and Insignia of the Royal Australian Air Force 1921–2018

John D. R. Macdonald

© Copyright 2025 Hugh Macdonald. No part of this publication may be reproduced, stored in a retrieval system, or transmitted in any form or by any means, including photocopying, recording, or other electronic or mechanical methods, without the prior written permission of the publisher, except in the case of brief quotations embodied in critical reviews and certain other non-commercial uses permitted by copyright law.

ISBN: 978-1-958407-37-0 (Casebound hardcover)
ISBN: 978-1-958407-38-7 (Soft cover)

The Royal Australian Air Force supports the publication of this book and has granted permission for the use of its badge. The views and interpretations expressed are those of the author and are not necessarily those of the Royal Australian Air Force or the Department of Defence.

Elm Grove Publishing
San Antonio, Texas, USA
www.elmgrovepublishing.com

CONTENTS

FOREWORD vii
PREFACE ix
ACKNOWLEDGEMENTS x

CHAPTER 1: THE BEGINNINGS OF AN AIR FORCE 1921 TO 1935   1

    Uniforms 2
    Rank Insignia 8
    Headdress 10
    Cap Badges 11
    Qualification Badges: Aircrew 14
    Qualification Badges: Ground Staff 20
    Other Badges 22
    Accoutrements 22

CHAPTER 2: CHANGE COMES TO THE RAAF 1935–1939   26

    Headdress 31
    Rank Insignia 34
    Qualification Badges: Aircrew 35
    Qualification Badges: Ground Staff 37
    Other Badges 38

CHAPTER 3: THE RAAF GOES TO WAR   40

    Working Dress 42
    Trainee Aircrew Dress 46
    Rank Insignia 49
    Headdress 53
    Qualification Badges: Aircrew 56
    Qualification Badges: Ground Staff 63
    Other Badges 66

CHAPTER 4: WOMEN'S SERVICES 1940–1977   74

    Royal Australian Air Force Nursing Service 74
    Women's Auxiliary Australian Air Force 85
    Women's Royal Australian Air Force 91

CHAPTER 5: THE POST–WAR PERIOD, KOREA & VIETNAM, 1948–1972   97

    Winter Service Dress 97
    Summer Service Dress 98
    Ceremonial Dress 99
    Tropical Dress 100
    Mess Dress 100
    Working Dress 101
    Rank Insignia 102
    Headdress 109
    Qualification Badges: Aircrew 113
    Qualification Badges: Ground Staff 119
    Other Badges and Insignia 119

CHAPTER 6: THE 'ALL SEASONS' UNIFORM 1972–2000 — 123

- Ceremonial Dress 128
- Tropical Dress 128
- Mess Dress 129
- Working Dress 130
- Rank Insignia 134
- Headdress 146
- Qualification Badges: Aircrew 152
- Qualification Badges: Ground Staff 155
- Other Badges 157
- Accoutrements 158
- Trial Patterns 162

CHAPTER 7: THE RAAF RETURNS TO DARK BLUE—2000 to 2016 — 164

- Working Dress 169
- Rank Insignia 171
- Headdress 172
- Qualification Badges: Aircrew 176
- Qualification Badges: Ground Staff 178
- Other Badges 182
- RAAF Uniforms in 2017 188

CHAPTER 8: RAAF BANDS — 189

- The First Band Uniforms 189
- Badges 193

CHAPTER 9: CHAPLAINS — 194

- First Chaplains' Uniform 194
- The 1930s 194
- World War II 196
- Post-War 197
- All Seasons Uniform 198
- Jewish Chaplains 199

APPENDICES 200

BIBLIOGRAPHY 214

JOHN MACDONALD 216

POSTSCRIPT 217

LIST OF APPENDICES

1. Rank Insignia – Commissioned Officers – Mid 1920s 201
2. Rank Insignia – Warrant Officers, Non-Commissioned Officers and Airmen – 1921 202
3. Uniforms – Commissioned Officers & Warrant Officers Class I – 1921 202
4. Uniforms – Warrant Officers Class 2, Flight Sergeants and Sergeants – 1921 206
5. Uniforms – Corporals and Airmen – 1921 207
6. Rank Insignia – Warrant Officers and Airmen – 1937 208
7. Uniforms – Commissioned Officers & Warrant Officers Class I – 1937 209
8. Uniforms – Airmen – 1937 210
9. Ranks and Special Insignia – 2015 211
10. Identified Manufacturers of Sterling Silver Flying Badges 212

FOREWORD

The origins of what we consider to be modern-day military uniforms can be traced back to 1337 and King Edward III who by all reports was shocked by the 'rag-tag' look of a body of Welsh spearmen raised to serve him abroad. He ensured each man was provided with a red tunic and mantle, thereby creating what was considered the first British military uniform.

Over the ensuing 800 years, uniforms have emerged and evolved in all forms and designs, from the flamboyant and at times highly impractical styles worn by many European armies throughout the 1800s and early 1900s, to khaki-coloured items and, more recently, camouflage patterned ensembles. What for some time was a means to visibly distinguish soldiers of one side from their enemies while on the battlefield in hand-to-hand combat has increasingly moved to one of utilising science and camouflage techniques to enable forces to be disguised and hidden from the enemy.

Today, modern uniform designs fulfil a range of functions. In addition to their utility in combat, they also enable the members of any military service to be readily identifiable amongst the community to which they serve. Uniforms can identify the country that a member serves as well as their respective Service. They can also inculcate a sense of pride and esprit de corps in the wearer of the uniform towards their nation and their Service.

While a comparatively young Service compared to the Royal Australian Navy, the Australian Army and their respective forebears, the Royal Australian Air Force has had considerable experience over the past 100 years in the evolution of its own uniform. While still possessing the obvious links back to the United Kingdom—the so-called 'mother country'—and the uniform designs of the Royal Air Force, the women and men of today's Royal Australian Air Force are readily recognised the world over as being Australian by the uniform they wear and the various accoutrements. The highly distinctive 'navy blue less one dip' colour selected by Sir Richard Williams back in 1921-22 for the uniform colour of the Royal Australian Air Force has greatly aided in achieving this recognition.

This book, *Shades of Blue,* written by the late John Macdonald, provides an extensive compendium of the uniforms worn by the personnel of the Royal Australian Air Force over these past 100 years. In its original form this work was a prize winner in the Air Force's 2016 Heritage Awards. Between 2016 and prior to his passing in mid-2018, John continued to work on the book to what you see before you today—his desire was to encapsulate the widest possible breadth of the numerous uniform types worn by the Royal Australian Air Force personnel as well as the various changes made along the way.

I consider myself very fortunate to have met John before his untimely passing and I have been constantly amazed at his knowledge of the subject. His spirit and endurance in those difficult days were exceptional. I am extremely honoured to provide this foreword to this book, and I congratulate John Macdonald's son, Hugh, for his efforts to have it published. It is a fitting tribute to a man who had a passion for military memorabilia and worked tirelessly up until his untimely death to ensure the book reflected as accurately as possible the breadth and complexity of the Air Force's uniforms over 100 years. With 2021 being the centenary of the Royal Australian Air Force, I can think of no better time for such a book to be made accessible to the public.

John Meier Air Commodore (Ret'd),
Director General of History and Heritage - Air Force
from March 2016 to May 2022

July 2021

PREFACE

This book started as a draft submitted to the Office of Air Force History as an entry in the RAAF Heritage Awards competition in 2016. Although it won second prize, it was felt that some additional research and editing would make it a more useful book for non-experts in this field. Wing Commander Terry Curtain (Retd) has done extensive archival research to clarify some points and dates, Squadron Leader Dave Burns (Retd) has reformatted and restructured the draft and Flight Sergeant Daryll Fell has worked his magic with the images. My sincere thanks to all of them.

Shades of Blue is the first known attempt at documenting the history of RAAF uniforms over the almost 100 years since the service formed in 1921. Despite all the efforts of contributors, the book cannot cover every RAAF uniform or badge ever worn. There are just too many variations, many of which have been forgotten or lost in the mists of time. Good photographs of uniforms, particularly those worn in the early years and during wartime, have been hard to find. This book concentrates on the uniforms most commonly worn by RAAF members, both on parade and in the working environment of airmen and airwomen.

The time available for research has been a major factor limiting the scope of the book. The many unauthorised variations of uniforms and badges, particularly those in wartime, have only been minimally covered. The huge subject of honours, awards, medals and commendations has been covered by other books and has not been included in this text. Flying clothing is a complex technical subject that has also not been covered, mainly due to lack of time for adequate research.

The uniforms worn by females in the first five decades of the RAAF have not been included in the chapters on those decades for the good reason that females were not members of the RAAF until 1977. In the decades before 1977, the female services were separate to the RAAF and the uniforms worn by their members are covered in chapter 4. The uniforms of female RAAF members after 1977 are included in the appropriate chapters.

Lastly, this book is not a dress manual and does not attempt to specify exactly how RAAF uniforms were worn under every circumstance. Plain English has been used rather than RAAF terminology to make it easier to understand for the average reader. Non-service personnel can quickly become frustrated by terms such as 'service dress 3B' or 'belt, man's, webbing, 80 cm'.

Shades of Blue is an honest attempt to show RAAF uniforms worn over many decades and is based on many years of research. No history can be 100 per cent correct and some errors or omissions have probably occurred. We hope that people who can enlighten us with information or good quality photographs will contact us and allow us to correct those errors in future editions. Any comments or suggestions should be directed to the publisher.

I hope that this book goes some way towards answering the many questions that people have asked me about the history of RAAF uniforms.

John Macdonald   *Melbourne, VIC*
January 2018

ACKNOWLEDGEMENTS

The original draft of Shades of Blue in 2016 was a product of the efforts of many people. I would like to recognise here the contributions of the many who have given of their time, their collections, knowledge and resources. Critically, without them and their contributions, this book would never have gone beyond a dream.

In particular, thanks go to my good friend Ian Jenkins,* with whom I have enjoyed a significant contact not only in my visits to Canberra, but also with lengthy telephone calls and emails in seeking information that defined and expanded my knowledge base in respect of the 1920s, 1930s and World War II periods, and particularly on the Empire Air Training Scheme. Access to his private collection has been critical in presenting a balanced picture for the potential reader. Along the way we have enjoyed the Hellenic Club and other venues around Canberra, as we have worked through the many drafts of the manuscript.

I cannot let the efforts of my editor, John Murdoch, go unrecognised. While John lives in the United Kingdom, there was much merit in having him at arm's length from the day to day activities. His expertise not only as an editor but also as a collector of Commonwealth Air Force insignia has been invaluable. John has also endured early morning telephone calls (to him at least) from me as I sought his counsel when an email just seemed inappropriate.

Without the input of Squadron Leader Steve Campbell-Wright I would never have gained access to copies of early documentation relating to uniforms and badges.

Thanks go to my dear cousin, Judith Williams OAM, who afforded me the comfort of accommodation and meals as I used her Yass, NSW home as my base for my research toils in Canberra. Time at Judith's place enabled me to step back and digest the impact of my research and discussions.

Thanks also to my valued colleagues, Lindsay Cheal, Dennis Graham and Robert Richards who have readily responded to my frequent calls for assistance in one form or another.

Thanks also go to the officers and warrant officers of the Royal Australian Air Force who have responded to the many queries that have been raised in the course of preparing this manuscript, and with apologies in advance if I have omitted any names as a result of the lengthy gestation period of this manuscript:

Air Vice-Marshal Tracy Smart, Air Commodore John Meier, Group Captain Jenny Fantini, Wing Commander (Retired) Terry Curtain, Wing Commander Steve Laredo, Wing Commander Harvey Reynolds, Warrant Officer of the Air Force Mark Pentreath, and Warrant Officers Geoff Banning and Richard Salcole.

The Directorate of History – Air Force (previously known as the Office of Air Force History) at Fairbairn, ACT has been most helpful and offered me encouragement in my efforts and, in particular, Martin James (Historian), Squadron Leader Dave Burns (Editor) and Steve Allan (Manager, Collections).

Also the RAAF Museum at Point Cook which, through David Gardner (Director), Emily Constantine (Collections Manager), Gary Walsh (Collections Curator), and Monica Walsh (Research Curator), has readily provided substantial assistance.

The Australian War Memorial, through its Curator Jane Peek and Manager Ron Schroer, has been most helpful.

There is another group of collector friends and colleagues, both here in Australia and overseas, who have readily assisted whenever they have been asked. Many have chosen not to have their names attributed directly to their contributions.

David Garner, John Marking, Mike Swan and Julian Tennant (Australia); Alex Bateman, Chris Morris and Mike Dresman (all of the United Kingdom); David Extance, Chris Langley, Ken Keegan and Mike McLean (all of Canada); along with Chris Kanca and Tod Rathbone (both of the United States of America).

Thank you also to my friend, Simon Floyd* who, with his diverse knowledge of Australian militaria was able to point me in the right direction when it seemed all bush and no sky.

When the complexities of Microsoft Word just stretched my patience, I was able to turn to my friend Sue Golding at Wonthaggi for assistance. Her husband, Brendan has been tolerant of my intrusions on their retirement from the work force.

Given the scattered geographic spread of the people involved in this project, it could not have been managed without the considerable technical knowledge of my eldest son, Hugh. His efforts have included building a website to enable the process of reviewing progress to be handled readily from one side of the world to the other. Hopefully by the time this manuscript reaches the publication stage, he will have received good news on the submission of his thesis for a Doctor of Philosophy.

I must mention my youngest son Andrew (aka Arnie) who died suddenly in the early days of this project, but who is in my thoughts daily. And last, but not least, my late wife Linda who initially encouraged me to undertake this project.

John Macdonald   *January 2016*

---

*Now deceased

Air Marshal Sir Richard Williams

CHAPTER 1:
THE BEGINNINGS OF AN AIR FORCE 1921 TO 1935

From 1914 to the end of 1919, Australian military aviation was the responsibility of the Army. At the end of World War I, the eight squadrons of the Australian Flying Corps (AFC) were disbanded when they returned to Australia in 1919, leaving Central Flying School (CFS) at Point Cook, Victoria as the only military flying unit in Australia. In 1919, the Australian Government decided to create an air service that was independent of the Army and Navy.

About the same time, the British Government made an offer of more than 100 aircraft, spare parts, tools and workshops, termed the 'Imperial Gift', to Australia for the purpose of setting up an air service. To receive this 'gift', the Australian Air Corps (AAC) was formed in 1920 as an interim air service.[1]

The AAC was duly established with Lieutenant Colonel Richard Williams (later Air Marshal Sir Richard Williams) as Director of Aviation Services. It took control of the former AFC training airfield at Point Cook, near Melbourne and the workshops and assets located there. Most of the AAC's personnel had served in the AFC. In addition to receiving and storing the Imperial Gift equipment, the AAC conducted flying operations to keep alive those military aviation skills that had been gained during World War I.

The AAC policy on uniforms was shown by the order issued by Captain H.H. ('Neil') Kilby, the Adjutant of CFS, less than three weeks after its formation:

...it has been approved that no distinct uniform shall be designed—that uniform at present in possession of members may be worn out, and that should members find it necessary to purchase new uniform while the Corps is constituted at present, AIF pattern, as worn by the AFC is to be worn.

Despite this order, over the next 12 months, the AAC gradually assumed an identity of its own. A major step in this direction occurred in November 1920, when the change was made from Army ranks to the air force-specific ranks that the Royal Air Force (RAF) had adopted in August 1919.[2]

On 15 February 1921, the Air Board recommended that the 'Australian Air Force' be formed as from 31 March 1921, and so Australia's third service was established. This was followed by the official announcement of the formation of the Australian Air Force in the Commonwealth of Australia Gazette No. 28 dated Thursday 31 March 1921. Later the same year, the Commonwealth of Australia Gazette No. 65 dated Thursday 18 August 1921 announced that Royal Assent had been granted for the use of the title 'Royal Australian Air Force' (RAAF).

Initially the RAAF comprised 21 officers, 128 airmen and 153 aircraft, which included 127 of the 128 'gift' aircraft from the British Government. Many of the personnel recruited for the new service came from the AAC, but not all.

---

[1] Air Power Development Centre, Pathfinder, *The Australian Air Corps*, issue 145, Canberra ACT, November 2010
[2] Ibid

UNIFORMS

Selecting the colour

As a new military service, the RAAF needed a uniform to replace the AFC uniform worn by members of the AAC. A design based on Royal Air Force (RAF) patterns was selected but the First Member of the Air Board, then-Wing Commander (later Air Marshal Sir Richard) Williams was not content with the colours being considered by the RAF. In recollection, Air Marshal Sir Richard Williams described the event in his autobiography.

"It was not possible for us to wait and see what the RAF was going to do before deciding on a uniform for our own Service, for it seemed to me to be essential that a new Service should have a distinctive uniform at the time of its formation. My idea was that our uniform should be blue, dark but readily distinguishable from the Navy. We had all likely warehouses in Melbourne and Sydney searched to see if we could find a suitable colour but without result; we could get navy blue and royal blue but nothing in between. Then one day I met the manager of the Commonwealth Woollen Mills at Geelong[3]. He asked me if I would care to look over his establishment and when I did, he was handling a batch of serge for the Navy.

This material was made up in natural wool and was then put through five machines in each of which it was dipped into indigo dye and when it came to the fourth machine (that is when it had had four dips) I found the colour I was looking for. We adopted the colour so it was in fact navy blue, less one dip."[4]

The first of the new blue uniforms began to appear in the second half of 1922, and it would not have been until 1923 that all personnel would have received their initial issue of the new blue uniform. In the interim, personnel wore previously issued Army attire.

Setting the Pattern: Forms of Dress

The uniforms of RAAF members were defined by the following forms of dress:

- Service dress was the equivalent military uniform to a business suit and was the most likely uniform to be worn when members appeared in public, for example at an official civil event or a military parade. It is also the dress that a member was most likely to be wearing when a studio photograph was taken. Service dress has summer and winter variants and is defined for officers, warrant officers, senior non-commissioned officers (SNCOs) and other ranks.
- Ceremonial dress was the uniform worn by members during a parade.
- Tropical dress was the uniform worn in tropical areas.
- Mess dress was a uniform worn to formal evening events in the mess and was worn only by officers, warrant officers and SNCOs. It is also known as 'mess kit'. The term 'mess undress' referred to a formal uniform that was equivalent to the civilian black-tie attire.
- Working dress is the everyday uniform worn while members are doing their daily duty. It includes variants for members engaged in aircraft maintenance, flying duties, warehouse duties or in medical sections/hospitals. Working dress uniforms are kept as simple as possible and typically show only rank insignia.
- Field service dress is the uniform when the member is working in a field environment and living in tented accommodation.

---

[3] The correct name was Federal Woollen Mills.
[4] R. Williams, These Are Facts, Canberra, The Australian War Memorial, 1977

Service Dress

The uniforms of the Australian Flying Corps in World War I were based on those of British Army cavalry units, with the trousers cut to be comfortable while riding a horse. The first service dress uniforms of the RAAF followed a similar style with a tunic style jacket and the trousers in the form of breeches or jodhpurs. All ranks below the rank of Wing Commander wore leather leggings.

Winter service dress tunic and breeches were made from a woollen serge material which was dyed dark blue. Gilt buttons displayed the eagle and crown. The summer equivalent was initially in white drill but when this was later found to be impractical, the colour was changed to khaki.

Officers' rank insignia was gold braid on the lower sleeve. Warrant officers, SNCOs and airmen wore their rank insignia on the upper sleeve. The officers' service dress rank braid changed later to black with a light blue central stripe, but the gold rank braid was retained on mess dress.

Service dress included all the qualification badges and insignia to which the member was entitled. Ribbons from campaign and gallantry medals were also worn. Aircrew flying badges were worn on the left breast above any ribbons.

Breast pocket button for a winter service dress jacket showing the eagle and crown

Front cover of 'Aircraft' published 31 March 1922 featuring the only known picture of the initial RAAF summer and winter service dress uniforms for airmen. Source: NLA Canberra

Winter Service Dress - Officers

Officer's winter service dress tunic, breeches & leggings from the early 1930s. Private collection

Summer Service Dress - Officers

Summer service dress consisted of khaki gaberdine tunic and breeches. In RAAF terminology, khaki coloured uniforms were commonly referred to as 'drabs'.

Tropical Dress

In June 1924, the Air Board approved the introduction of the first form of tropical dress which was to be worn by RAAF personnel proceeding to tropical areas to work with the Navy. The dress consisted of wool khaki shirt, khaki shorts, blue stockings (long socks) and canvas shoes. The dark blue stockings were later changed to khaki.

Officer's summer service dress tunic, circa 1928

*A group of air cadets in tropical dress at RAAF Richmond in 1933.
Source: T.F.C. Lawrence Collection*

Ceremonial Dress

For ceremonial occasions, officers wore their service dress uniform, embellished with a gilt sword belt and sword, and full-size medals as shown below.

RAAF Guard commander speaking with his Royal Highness the Duke of York (later to become King George VI) Sydney 1925.
Source: State Library New South Wales

Airmen wore service dress with a white belt and gold buckle replacing the blue belt of the tunic.

RAAF guard of honour in Melbourne in 1925. Source: RAAF Museum

Mess Dress

By the mid-1920s, RAAF officers were wearing a mess dress based on the one worn by the RAF. It consisted of a short, dark blue jacket with light blue silk facings. The trousers had leg stripes of light blue. An Eton-style waistcoat in dark blue with gilt buttons was worn under the jacket. Rank insignia consisted of gold braid stripes worn on shoulder straps surmounted by a gilt eagle and crown. A white, wing-pointed dress shirt and black bow tie were worn with mess dress. Officers qualified to wear a flying badge wore a miniature badge on the left lapel.

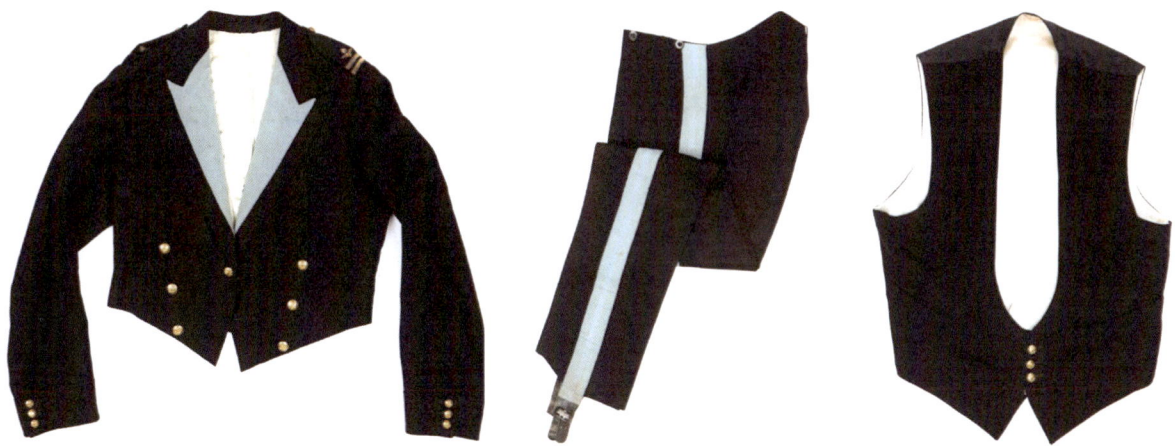

Mess dress jacket, trousers and waistcoat with light blue silk facings and gilt buttons.
Source: AWM Canberra

A tropical version of mess dress was in existence by 1931. It consisted of jacket and trousers of similar style, in white drill material. A vest of white Marcella material was worn under the jacket.

Working Dress: Officers

The officers' working uniform of the day was full service dress in winter and service dress without the tunic in summer. It was not appropriate to wear the winter service dress without the tunic because the white shirt did not have rank insignia or other badges. However, in the summer, the khaki shirt worn with summer service dress did have rank insignia on shoulder straps and so it was appropriate to wear summer service dress without a tunic.

Flying was often conducted in service dress and, if a flying suit was used, it was usually worn over the top.

Working Dress: Airmen

Airmen engaged in heavy duties such as aircraft maintenance and warehousing, wore dark blue overalls as working dress. A dark blue beret was worn with overalls. It had the advantage of being able to be stowed in the airman's pocket when he was crawling around the aircraft. Airmen employed in office duties usually wore the appropriate service dress when on duty.

RANK INSIGNIA

Like many other aspects of the RAAF uniform, the system of ranks and the rank insignia were derived from the RAF. For the origins of these rank names and insignia, see Hering, 1961.[5]

In November 1920, the Australian Air Corps, the predecessor of the RAAF, replaced its army ranks with air force ones so that when the RAAF was formed on 31 March 1921, the use of air force ranks was already well established.

Officer Rank Insignia

In 1921, the Air Force officer rank insignia were the same as those used by the RAF, which in turn were based on the system of rank braids used by the Royal Navy. Variations from the naval rank insignia were the use of a single thin braid for Pilot Officer and the introduction of a broad braid for air rank officers. On RAAF service dress uniforms, officer rank insignia consisted of gold braid sewn on the lower sleeves with opposing eagle and crown badges above.

Rank insignia for a Squadron Leader on a 1928 service dress uniform

Eagle and crown badges. Left arm (top) & right arm (bottom)

Non-Commissioned Rank

The RAAF non-commissioned rank badges were also derived from their RAF equivalents which were, in turn, based on those which had been used in the Royal Flying Corps. The new air force ranks were:

- sergeant major, class I (later known as warrant officer, class I)
- sergeant major, class II (later known as warrant officer, class II)
- flight sergeant
- sergeant
- corporal
- leading aircraftman
- aircraftman.

---

[5] *Customs and Traditions of the Royal Air Force*, Sq Ldr P G Hering, Gale & Polden 1961.

The chevrons worn by RAF sergeants and corporals were the same as those worn by their army equivalents but were embroidered in blue instead of khaki. The new flight sergeant rank included the crown of the RAF Flight Sergeant and the Army staff sergeant. The British coat-of-arms (with the lion and unicorn) was used for the RAAF warrant officer class 1 badge until 1973 when it was replaced with the Australian coat-of-arms containing the kangaroo and emu.

The insignia worn on service dress were as follows.

Sergeant Major class I / Warrant Officer class I

Warrant Officer Class II / Sergeant Major Class II

Flight Sergeant, Service Dress

Flight Sergeant, Working Dress

Flight Sergeant, Summer Service Dress

Sergeant, Service Dress

Sergeant, Working Dress

Sergeant, Summer Service Dress

The Flight Sergeant and Sergeant insignia were worn above three chevrons

Leading Aircraftman, Service Dress

Leading Aircraftman, Working Dress

Leading Aircraftman, Summer Service Dress

There were no rank insignia for aircraftmen.

HEADDRESS

Service dress

The headdress worn with service dress by all ranks was the blue peaked cap with a black braided band around it, as shown below. Badges and other embellishments varied with rank.

Pictured is a 1920s-1930s senior officer's service dress cap fitted with a second pattern officer's cap badge. Wattle leaves are embroidered on the peak, with single row being worn by officers of Squadron Leader to Group Captain rank. Officers below Squadron Leader and warrant officer class I wore no leaves on their cap peaks

The dark blue cap was worn with winter and summer service dress, mess dress and mess undress. When worn with a khaki uniform, a khaki colour cover was fitted over the peak cap, as shown

CAP BADGES

In his autobiography, Air Marshal Sir Richard Williams wrote:

'When it came to cap badges I called on the photographic officer I had in No 1 Squadron AFC, Lieutenant O H Coulson, who was quite an artist, and he designed them. I wanted to introduce the wattle into these badges, and it was quite successful in those for other ranks in brass. In the case of officers, however, while the design was suitable in appearance we had great difficulty in getting it produced in gold embroidery; each batch of badges turned out larger than the last and we later adopted the RAF Badges. We used the same eagle in gilt as did the RAF in the officers' badges until we came to consider the standard design for the mottoes of the Service and the individual units. We then realised that, in the wedge-tailed eagle, Australia had one of the largest in the world and we adopted it. The College of Heralds did not know much about this eagle and there was trouble getting the badge right.

Recently a gilt wedge-tailed eagle has been introduced into the officer's cap badge to its considerable improvement.[6]

The cap badges for officers and airmen that accompanied the introduction of the new uniform were also uniquely Australian. They featured full wreaths of wattle leaves, interspersed with golden wattle balls and surmounted by a Tudor Crown with the red cap of maintenance (often referred to as a red cushion) in the crown.

Cap Badge — Officers

The original 1921 document described the badge for officers and warrant officers class I as follows: 'Wreath of wattle surmounted by a crown—all in gold embroidery. Southern Cross embroidered in gold on blue ground inside wreath. Eagle in gilt metal superimposed.'

It was soon realised that this first pattern was astronomically incorrect as the Southern Cross cannot be seen during the day, when a blue sky is visible, and is only ever seen against the night sky which is black. So, by 1927, a second pattern of this badge was produced with the background material changed to black velvet.

---

[6] *These are Facts,* Richard Williams, The Australian War Memorial, 1977

First Pattern

Mystery has long been associated with the existing first pattern badges, especially given their short period of usage by the RAAF, and it was thought only two examples survived, one of which was in a poor condition. The one accessible badge had been restored, most probably in the 1950s–1960s, by the addition of a replacement cast, gilt eagle. The backcloth on the reverse had also been replaced after fitting the new eagle. Otherwise, the obverse is a genuine and virtually mint example of this first RAAF service dress cap badge. This badge was used as a pattern for a limited number of high quality reproduction badges in 2015 which were sold in pairs with eagles flying in both directions. They can easily be identified by the lower two stars being in the 'correct' angled position and the striped nature of the light blue background.

Second Pattern

Confirmed in the 1931 Dress Regulations (which replaced the 1929 edition) with the background to the Southern Cross being black velvet (Appendix 5/B/2). This second pattern badge has the eagle in sinister (left facing) mode, but some variants of the badge had the eagle in the dexter (right facing).

Third Pattern

In the mid-1930s, a third pattern badge was introduced. On this badge, the leaves were joined without a knot at the base of the wreath and the lower stars have been separated. As these badges were hand-made, many variations occurred between individual badges.

First pattern Officers' and Warrant Officers' cap badge   *Second pattern cap badge with eagle in sinister mode [7]*   *Third pattern cap badge*

Some makers' labels on the reverse of badges are shown opposite. These three recognised patterns of the officer's service dress cap badge resulted in six variations as they were made with the eagle flying in either direction. However, the hand-made method of producing the cap badges brought with it problems. Air Marshal Sir Richard Williams wrote:

'In the case of the officers, however, while the design was suitable in appearance, we had great difficulty in getting it produced in gold embroidery; each batch of badges turned out larger than the last and we later adopted the RAF badges' [8]

---

[7]   The direction of the eagle is determined by the direction in which it appears on the wearer's uniform or cap—to his left is sinister and to his right is dexter.

[8]   *These are Facts,* R Williams KBE CB DSO, Australian War Memorial and the Australian Government Publishing Service, 1977 ISBN 0 642993998 p. 136

I Montgomery, Melbourne *David Jones, Sydney* *Harvey Smith, Sydney*
Second pattern, above *A third pattern badge* *Third pattern, above*

Cap Badge, Airmen

The cap badge for airmen (warrant officer class II and all ranks below) was in pressed gilt brass finish, smaller, but in a similar style to that of the officers. It again featured the full wreath of wattle leaves, interspersed with wattle balls and surmounted by a Tudor Crown, but with the central monogrammed letters 'RAAF' instead of the Southern Cross and eagle. The design of the airmen's badge, except for a change of crown, has remained unaltered since its introduction into the RAAF in 1921. It also forms the basic central motif of the pilot's brevet.

Wattle Leaves on the Cap Peak

Officers of the rank of Squadron Leader, Wing Commander and Group Captain wore a single row of gold wattle leaves on the peak of their service dress caps. Officers of air rank (Air Commodore to Marshal of the RAAF) wore two rows of gold wattle leaves on the peak of their service dress caps. Junior officers (Flight Lieutenant, Flying Officer and Pilot Officer), warrant officers and airmen wore no wattle leaves on their caps.

The peak of a service dress cap, with one row of wattle leaves, as worn by senior officers

The peak of a service dress cap, with two rows of wattle leaves, as worn by air officers

Air Cadets Hat Band

From the time of the first course training new pilots in 1923, trainees joining the RAAF direct from civilian life were given the rank of air cadet. Air cadets were identified by a white band on their service dress cap in place of the black band worn by other RAAF members. Cadets wore officers' service dress uniform with no rank braid but with the eagle and crown badge on each sleeve.

Helmet, Wolseley Pattern

The Wolseley pattern helmet was also variously known as topee, solar topee and drab helmet. It is a distinctive British design developed and popularised in the late-19th/early-20th century. With its swept-back brim, it provided protection from the sun and its use was soon widespread among British personnel serving in tropical areas.

The helmet first appeared in RAAF Document No. 17, Priced Vocabulary of Clothing and Necessaries, dated 22 July 1929. It remained listed until 1941 and was withdrawn during the course of World War II, with the khaki fur-felt hat becoming the preferred headwear.

The helmet was worn with tropical dress. Around the helmet was a band of folded khaki material called a puggaree. The fourth fold was dark blue in colour. In the late 1930s, the blue fold was removed from the puggaree.

 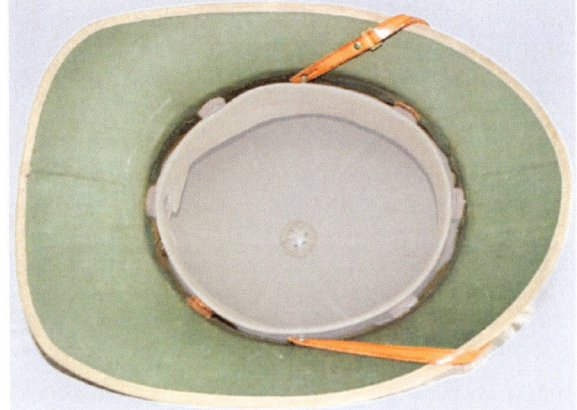

Helmet, Drab with airman's cap badge, puggaree with blue fold, and brown leather chinstrap

QUALIFICATION BADGES: AIRCREW

Pilot's Flying Badge

The badges awarded to military aircrew on their graduation from flying training are commonly known as 'brevets'. The first pilot brevet in the world was that designed for pilots of the Royal Flying Corps (RFC) in the British Army. The initial design was undertaken by two British Army officers who had started 'doodling' on the blotting paper during a Royal Flying Corps meeting. The resultant proposal was submitted to King George V who approved the design in 1913.[9]

The badge consisted of the letters 'RFC' surrounded by a wreath and surmounted by a Tudor crown with an outstretched wing on either side. The approval documents do not state which bird was used in the design, but the War Office file on the initiation of the badge contains an illustration of the wings of a swift. It is likely that this influenced the final design of the RFC badge.[10] When the RAF was formed in 1918, their pilot brevet was identical to the RFC badge but with the monogram 'RAF' in place of 'RFC'. The RFC design was also adopted by the Australian Flying Corps, with a central 'AFC' monogram and was subsequently adopted by the RAAF, with an 'RAAF' monogram.

---

[9] Squadron Leader Peter Hering, *Customs and Traditions of the Royal Air Force*, Gale and Polden, Aldershot Hampshire UK, 1961, p 101

[10] Ibid p.104

The first pattern RAAF pilot brevet was introduced in 1921 for wear on the khaki service dress tunic. It was also worn on the flying suit and summer uniform tunic.

The main features of the brevet are drab silk embroidered wings, crown and monogram, enclosed in a brown wreath, on a black wool patch. Other patterns of brown-wreathed RAAF pilot brevets were worn well into the mid-1930s, but these were embroidered on khaki background material instead of black material. This brevet is particularly rare.

The first pilot brevet was made for wearing on the khaki tunic

When the dark blue uniform was introduced in 1922, a gold-on-blue brevet was created. The main features of this brevet were the golden-yellow wings, crown, wreath and lettering, with gold wire embellishment around the feathers and central wreath. It was embroidered on dark-blue uniform material. The central lettering was monogrammed in the form of the airman's cap badge, while the wing was styled along the leading edges similar to the 1918 AFC pattern, but less pronounced than its brown-wreathed counterpart.

The RAAF gold-on-blue pilot brevet was produced for the new service for wear on the dark blue service dress uniform, worn here by Flying Officer U. E. Ewart in July 1924.

Bullion Flying Badges

By late 1924, the first bullion wings appeared. The hand-embroidered gold service dress pilot brevets of the 1920s and 1930s underwent significant style changes in crowns, wreaths, wings and monograms. There were variations in the style of the Tudor crown, some with the red velvet cap protruding well outside the base, but always with a nicely curved top, instead of the 'hip-roofed' style of the reproductions. Wreaths varied from having many leaves to only a few, as did the ties at the base of the wreath. Wings varied from the 'AFC' style in the early 1920s; to nicely curved with flat-base feathers in the mid period of the decade; and to the elongated style of the latter half of the decade (examples were those worn by record-breaking aviators Charles Kingsford-Smith and Charles Ulm in 1928). By the middle of the following decade, they had progressed to a wing with a curved leading edge and flat base feathers. Monograms varied from the style of the airman's cap badge to rather rough designs in both silver and gold bullion.

1924 pattern bullion brevet

1927 pattern bullion brevet

Pilot's Flying Badge Type A

RAAF clothing documents from 1929 listed two types of pilot brevet - 'Pilot Gold on Blue' and 'Pilot Gold on Blue "A"'. The brevet 'Pilot Gold on Blue' was also known as the 'general flying badge'. The difference between these brevets remained in place until 1940 when paragraph two of ABO N.515 dated 30 August 1940 stated that: 'Officers who were previously entitled to wear flying badge 'A' may now wear the approved 'general flying badge' with prescribed orders of dress'.

To date, no documentary or photographic evidence has surfaced which explains the operational differences of the two types of flying badges. However, at this time, the RAAF conducted two flying courses—one of eight month's length aimed for permanent officers and a shorter course of five months for Citizen Air Force officers. It is possible that the two different brevets distinguished between the graduates of the two pilot courses.

White-on-Blue Flying Badge

By 1928, the embroidered brevet had changed to white-on-blue and was commonly worn on the khaki summer tunic or khaki shirt. The brevet was hand-embroidered in white thread on RAAF blue background material. The feathers on the wings were picked out in black thread, while the crown had a red 'cap of maintenance'.

Change from Brown to Blue Wreath

About the mid-1930s, the embroidered pilot brevet was changed to one with the wreath in blue instead of white or brown. This was a significant change from the brown wreath of the RAF brevet.

1934 pattern white-on-blue pilot brevet

Observers

During World War I, observers were officers carried in aircraft to conduct reconnaissance. Initially, observers, who shared the dangers of battle with their pilots, had no insignia to identify them as combatant fliers until September 1915 when the Royal Flying Corps introduced the observer brevet. The brevet consisted of a single wing protruding from the letter 'O' embroidered in white silk. It used a different wing to the RFC pilot brevet, without the crown or the monogram.. To identify its observers, the AFC used a similar brevet,.

When the RAAF was formed in 1921, the small number of members who had qualified as observers in the AFC were entitled to wear the observer brevet. During World War II, the category of 'observer' was changed to 'navigator' and the observer brevet evolved into the navigator brevet.

From documents, two patterns of the observer brevet were worn in the early years of the RAAF. The most common one was a copy of that worn by the AFC. It was an embroidered badge of white thread on black background material.

Observer brevet similar to the one worn by AFC observers.
Source: Australian War Memorial, REL37431

The only known photograph of the second pattern observer brevet, which incorporated the RAAF monogram within the 'O', is shown below. The photograph, taken in 1924, shows Flying Officer W.S.J. Walne who was the only observer officer commissioned into the RAAF at its inception in 1921.

Flying Badges—Working Dress

Working dress flying badges for Pilots and Observers can be seen on uniforms in early 1923 and appear to have been embroidered in a thread colour that is very similar to the background colour, which itself closely matches that of the uniform. Some of these pilot flying badges have a very light-coloured wreath, while others do not. The same thread and background colours apply to the observer flying badge, while the large 'O' in which the RAAF monogram appears, is embroidered in the same light colour as the pilot's wreath. The reason for these visual differences is unknown.

Drab Flying Badges

By the early 1930s, the RAAF introduced 'drab' flying badges for pilot and observer aircrew categories for wear on drab uniforms. Little is known about these badges and examples are extremely rare, but photographic evidence clearly supports their existence.

The flying badge had embroidered beige-coloured wings, crown and 'RAAF' monogram inside a brown wreath on a khaki background material which is slightly darker than the jacket material. The Tudor crown has a rather 'chunky' appearance, and the wings are flat-based.

Air Gunner Badge

Three airmen completed the first Air Gunner's course at No. 1 Squadron, Laverton in June 1930 and became entitled to wear the 'winged bullet' air gunner's badge. This was initially worn immediately above the rank badge on the right sleeve but, from 1931, it was worn on both sleeves. The role of air gunner was performed as a secondary role to the Airman's principal mustering.

The brass badge was a copy of the RAF badge introduced in 1923 and the initial badges were probably sourced from the RAF. This version had three lugs with no rim at the base of the bullet. Badges with two lugs and badges with a rim are commonly found and it remains a matter of debate whether these versions were official issues or later copies. The consensus seems to be that badges were issued with either two or three lugs, but the provenance of badges with a rim is less certain. There is no record of any variation of this badge having a maker's name.

 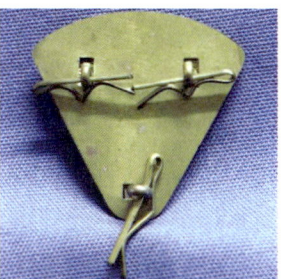  

Air gunner badge (left to right) with 3 lugs, backing plate, two lugs and rim

Governor of New South Wales, Sir Philip Game, speaking with Drum Major 2940 Patrick Stanislaus Norris wearing the air gunner badge above his rank chevrons at Government House, Sydney, 1934

---

[11] P.G. Hering, *Customs and Traditions of the Royal Air Force*, Gale and Polden Limited, Aldershot, 1961 p.109

QUALIFICATION BADGES: GROUND STAFF

Medical Services

The initial badge introduced in the 1920s for medical personnel was the Geneva or Red Cross armband which was worn by both medical officers and trained nurses (male). The latter were the Air Force equivalent of the Army stretcher bearer and were drawn from the non-commissioned ranks. A Geneva Cross badge embroidered on a white circle and backed on to blue uniform material was also worn on winter dress, and a similar badge on drab uniform material for summer dress.

Trumpeter

The embroidered trumpeter badge was introduced in the 1920s for trumpeters and buglers, embroidered in blue for winter dress and drab for summer dress. It was worn on the sleeve, immediately above rank insignia. The badge was no longer in use by 1931.

Trumpeter badge for winter dress

Wireless/Telegraphist Operators and Operator Mechanics

The wireless/telegraphist (W/T) operator and operator mechanic badge was brought into RAAF use in the early 1920s to recognise the skills of the first electronic communications members. The badge was described as a 'winged thunderbolt' and was embroidered in light blue and white silk on a dark blue background. It was worn immediately below the eagle badge on both sleeves of the jacket. On the summer uniform, the badge was on drab material.

1920s W/T operators & operator mechanic *Army-trained W/T operator/ mechanic*

With the rapid development of wireless (radio) communications during this period, some of the Wireless/Telegraphist training was conducted by the Army. Army personnel seconded to the RAAF wore RAAF uniforms but retained their appropriate Army trade badges during their period of secondment.

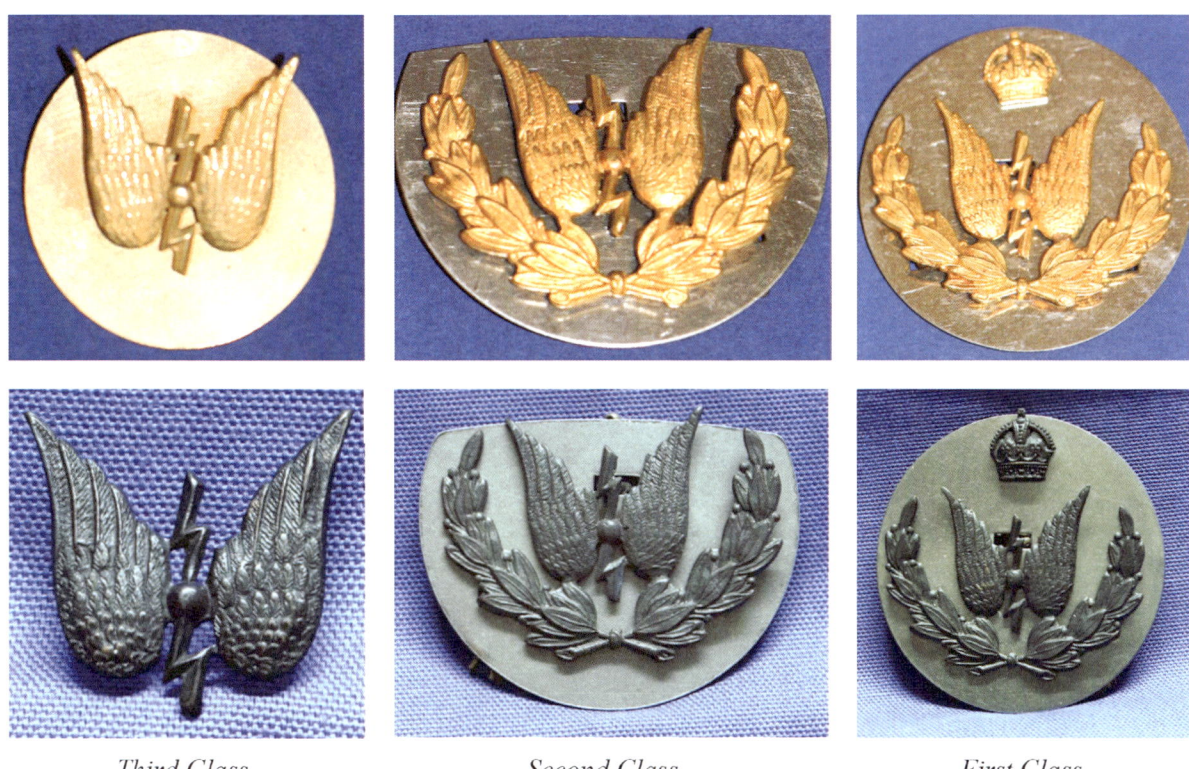

Third Class *Second Class* *First Class*

Army Wireless/Telegraphist Operator/Mechanic qualification badges.
(The oxidised versions are from WWII)

This practice continued into the Second World war as witnessed by the illustrated WWII khaki cotton drill jacket made by MTX which has 4 black plastic RAAF buttons, with smaller ones on the two top pockets. It has a belt hook on the left side and the belt is held in place by loops and a plastic buckle. It has sergeant's chevrons on both sleeves and an oxidised Army-trained wireless/telegraphist qualification badge on the right sleeve. It is named inside and on the reverse of the belt "115052. Ronald Paul Mills enlisted on 16 July 1942 at Adelaide and discharged as a Sergeant on 8 Jan 1946 from Air Defence Headquarters Townsville".

OTHER BADGES

Shoulder Badges

Eagle badges were worn in opposing pairs, flying to the rear, one inch below shoulder seams, by all ranks up to Warrant Officers Class II. Initially embroidered in light blue silk on a dark blue background for service dress, the pattern changed over time with the introduction of variations for working dress and the summer uniform. Officers and Warrant Officers Class I wore the gilt eagle and crown badge on the lower sleeves. See p38 for later patterns.

First pattern: Service Dress

First pattern: Working Dress *Second pattern: Working Dress* *Second pattern: Khaki Drab*

Citizen Air Force

In 1925, the RAAF established the Citizen Air Force (CAF) which allowed members to serve on a part-time basis. CAF personnel were identified by the triangular CAF badge which was based on the AFC colour patch worn during World War I. This badge continued in use, with variations, until the CAF was re-organised into the RAAF Active Reserve in the 1980s, when the badge disappeared from use. The badge was worn one inch below the shoulder seams on each arm.

Citizen Air Force sleeve badge

ACCOUTREMENTS

Swords

Officers and warrant officers in ceremonial dress commonly carried swords. The RAAF sword was a copy of that introduced by the RAF in 1925. The sword had a single-edged straight blade with gold-plated brass hilt, white fish-skin grip and a brass pommel in the form of the head of an eagle. A stamped gold-plated brass cartouche bore the eagle and crown emblem. The scabbard was made of rawhide with gold-plated brass mounts and the scabbard for Air Rank Officers was heavily chased. Air Rank Officers also had a different buckle on the slings.

The RAAF sword has undergone few changes since it was introduced in the 1920s. The major change has been the change of the royal cypher when a new monarch was crowned.

Brassards

By 1931, brassards were worn by the following duty personnel on the right arm above the elbow:

- Duty Pilot[12] and Non-Commissioned Officers of the Watch wore a red cloth brassard.
- Service Police members on duty wore a black brassard with red cloth letters 'S.P.' ¾ inch in height.
- The Orderly Officer and Orderly Non-Commissioned Officer wore a light blue brassard with a dark blue stripe ¾ inch wide in the centre.

Officers of No 101 Flight at RAAF Station Richmond NSW in April 1932. The officer at the far right is wearing the brassard of an Orderly Officer

Orderly Officer brassard in use between 1925 and 1940

---

[12] The role of the Duty Pilot was to control the movement of aircraft within the air base, effectively the air traffic function of the day.

Ceremonial Belt Buckles

During this period the RAAF had two different buckles, one for wear by officers on their sword belt and another by airmen on a white leather belt. Both of these were worn on ceremonial parades.

Officer's Ceremonial Belt Buckle

Airman's Ceremonial Belt Buckle

Branch Identification on Officers' Rank Braid

Following the Royal Naval tradition of identifying branches of service, the fledgling Air Force made provision for non-flying personnel to have a distinguishing colour of material between the rows of rank braids on uniforms and shoulder straps. The designated colours were:

- Officers on Quartermaster's List other than Technical Officers—WHITE
- Technical Officers—PURPLE
- Medical Officers—RED

The requirement was still recorded in Appendix 5/B/1 published on 22 June 1931 but had disappeared by the time of the next publication on 5 July 1937.

Manufacture of Officer Uniforms

While officers' uniforms were readily available through RAAF sources, from advertisements of the period it seems officers could have uniforms privately tailored and the various materials appeared in Publication No. 17 with prices per yard (imperial measurement). An advertisement for a tailor who specialised in making RAAF officer uniforms in 1921 is shown here.

The Image of the RAAF was Established

By 1935, the RAAF's uniforms had matured into a set of uniforms that identified members of the new service and that were suitable for the range of activities under the climatic conditions likely to be encountered. Dark blue was established as the colour of the Air Force. The photograph below shows the image that the RAAF projected to the Australian public at this time.

Air cadets in service dress at Point Cook, VIC circa 1935

CHAPTER 2:
CHANGE COMES TO THE RAAF 1935–1939

UNIFORMS

By the middle of the 1930s, two major events would have a profound effect on the development of the RAAF. Firstly, German rearmament in Europe and Japanese expansionist policies in Asia forced the Australian Government to face the imminent threat of war.

Secondly, Australia and the rest of the western world were recovering from the Great Depression. A healthier economy made funds available to expand Australia's defence capabilities, including a major expansion of the Air Force.

A 1928 report by Air Marshal Sir John Salmond, a senior British officer, had recommended a major expansion of the RAAF. This report had not been acted on immediately, due to lack of finances, but by 1935 it became the template for the expansion.[13] The RAAF went from a service of 956 officers and men in April 1935, to one of 3,172 members by April 1939, a three-fold increase over four years.[14] With the increased budget came orders for modern warplanes, including Bristol Beaufort bombers, Lockheed Hudson bomber/transports and Short Sunderland long-range maritime patrol aircraft (the RAAF's first four-engined aircraft). With the threat of war, the RAAF transitioned from a flying club mentality to that of a professional military force.[15] Gone was the image of the RAAF pilot in his riding jodhpurs sitting in the open cockpit of his biplane with his scarf flying in the wind.

With the more professional outlook came major changes to the RAAF uniform. The Salmond Report of 1928 had recommended greater standardisation between the RAAF and the Royal Air Force (RAF) in equipment and training. This was also to flow through to the uniforms and insignia worn by members of the service, although there was to be no wavering from the original choice of basic uniform colour—it remained dark blue.

In addition to changes to the existing forms of dress, two new forms of dress were added— full dress and field service dress. The changes are detailed in the following paragraphs.

Full Dress

The concept of the full dress uniform was that it equated to the civilian top-hat and tailcoat and it gave Air Force officers an equivalent uniform to that worn by their Navy and Army counterparts at state and vice-regal occasions. The RAAF full dress uniform for officers was approved in 1936.[16] The winter full dress was based on the RAF full dress but in RAAF's dark blue colour. It was also to be at no cost to public funds; in other words, officers were to purchase their own uniforms. As this uniform was costly to purchase, very few officers actually owned one. It also had a very short life as it was withdrawn from use on 21 November 1939[17]

---

[13] Alan Stephens, *The Australian Centenary History of Defence, Vol II, The Royal Australian Air Force,* Oxford University Press, Melbourne VIC, 2001, p 53

[14] Coulthard-Clark, C D, *The Third Brother – The Royal Australian Air Force 1921-39,* Allen and Unwin, North Sydney NSW, 1991, appendix 4

[15] Alan Stephens. As footnote 13

[16] Air Board Agenda No. 1853 of 21 February 1936

[17] Air Board Order N.230 Authorised Orders of Dress Para 4 [1]

for the duration of World War II and, in line with its RAF counterpart, it never re-appeared after the war.

Rank was indicated by gold rank lace on the sleeve and, additionally, by oak leaves and acorns in gold embroidery on the collar as follows:

- Pilot Officer to Flight Lieutenant   2 acorns
- Squadron Leader to Group Captain   5 acorns
- Air Commodore to Air Chief Marshal   13 acorns

An embroidered eagle and crown badge was worn on the epaulette on each shoulder. For officers of air rank (Air Commodore to Air Chief Marshal), the button securing the epaulette to the jacket was surrounded by a laurel wreath.

Air Vice-Marshal Sir Richard Williams, in winter full dress uniform.
This uniform is part of the collection within the RAAF Museum, Point Cook

A tropical version of full dress was described in RAAF regulations. It consisted of the winter full dress but with a white tunic, white trousers and a white busby. However, as the RAAF did not have bases in tropical areas at the time, it is likely that the tropical full dress was never worn. Certainly, no photographic evidence of officers wearing the tropical version has been located.

---

[18] *Customs and Traditions of the Royal Air Force*, P G Hering, Aldershot Gale and Polden 1961 p. 224

Service Dress

The service dress described in chapter 1 continued in use with the following changes.

- By 1937, the wearing of breeches, leggings and puttees by officers and airmen had ceased. All ranks wore dark blue trousers for winter service dress or khaki trousers for summer. Boots were no longer worn by officers and warrant officers class I, who wore black leather shoes with all uniforms. Airmen of flight sergeant rank and below continued to wear black leather ankle boots or shoes.
- With the introduction of field service dress as the everyday working dress for most members, service dress was only worn on special occasions such as a military parade.
- Some changes also occurred to airmen's rank badges—see section below on rank insignia.
- New badges were introduced on service dress caps for officers and warrant officers—see section below on headdress.

Probably the arrival in England of the RAAF contingent for the 1937 Coronation in May 1937. Note three sergeants still wearing a propellor over chevrons & an LAC with the air gunner badge

Mess Dress and Mess Undress

By the late 1930s, the design of officers' mess dress was aligned to that of the RAF. The same style of dark blue jacket was worn but the gold braid rank stripes had moved to the lower sleeve in a similar position to winter service dress. Under the jacket was worn a white waistcoat, white dress shirt and black bow tie. The trousers remained dark blue but the light blue stripe had disappeared.

The tropical version of mess dress was similar in style to the winter version but the jacket was white instead of blue. Either white or dark blue trousers could be worn with this version. Rank insignia remained on the epaulettes. Winter mess undress was similar to mess dress except that the waistcoat was blue instead of white. A tropical mess undress included the option of wearing a blue silk cummerbund instead of a waistcoat.

Field Service Dress

By 1937, the RAAF introduced field service dress which was a more practical uniform for wear in a working environment or when deployed on operations or exercises.

RAAF Guard on exercise Western NSW 1937

Winter field service dress for both officers and airmen was based on winter service dress but with black buttons and belt buckle.

Any badges worn were embroidered or of black oxidised metal finish.[19] Officer rank braid was still on the lower sleeve but in black lace with a blue stripe in the centre. A summer version of field service dress consisted of summer service dress but with black buttons, oxidised badges and the black/blue rank braid.

The field service cap (see section below on headdress) was introduced to be the headdress worn with both summer and winter field service dress. Officers and warrant officers class I had the option of wearing a service dress cap instead of the field service cap with field service dress.

Even though the introduction of field service dress provided airmen with a more practical working uniform, most airmen employed on aircraft maintenance duties usually wore dark blue overalls and dark blue beret.

---

[19] This is the first mention of oxidised badges in the RAAF dress regulations.

Working dress: dark blue overalls and dark blue beret. Source: State Library Victoria

Tropical Dress

Tropical dress of khaki short sleeve shirt and khaki shorts continued to be worn by RAAF members working in tropical areas. During the late 1930s, the colour of the stockings (long socks) worn with this dress changed from blue to khaki. The Wolseley Pattern helmet was the standard headdress worn with tropical dress, but the field service cap could also be worn.

Tropical dress RAAF Amberley 1937

HEADDRESS

Full Dress Busby

The full-dress busby and an enlargement of the gilt badge with silver eagle which was only worn on this headdress

Service Cap

The dark blue service cap with black mohair hatband worn in the 1920s continued to be worn by all ranks through the late 1930s and the badges for Aircraftmen to Warrant Officer Class II and for Warrant Officer Class I remained the same.

Airman to Warrant Officer Class II

Warrant Officer Class I

Officers from Pilot Officer to Group Captain

The cap badges worn by officers and warrant officers class I in the 1920s and early 1930s proved difficult to manufacture with any consistency.[20] In 1936, the decision was taken to adopt the cap badges used by the RAF for officers and warrant officers class I.

This badge is a good example of the one worn by officers from Pilot Officer to Group Captain, although there was some variation in the shape of the background material. This badge is described as an eagle in gilding metal above entwined laurel leaves of gold embroidery, the whole surmounted by a crown of gold embroidery.

In addition to the badge, Group Captains wore one row of gold oak leaves on the patent leather peak of the service cap. Officers from Pilot Officer to Wing Commander had a cloth peak without any embellishment.

Officers from Air Commodore to Air Chief Marshal

Air rank officers wore a service dress cap with a black patent leather peak and a gold bullion badge. The badge was described as a wreath of laurels surmounted by lion and crown, all in gold embroidery. An eagle in gilding metal superimposed over the wreath. In addition to this badge, air rank officers wore two rows of gold oak leaves on the patent leather peak of the cap.

---

[20] Williams, Sir Richard, Air Marshal (Retd), *These are Facts – The Autobiography of Air Marshal Sir Richard Williams, KBE, CB, DSO*, Australian War Memorial and Australian Government Publishing Service, Canberra ACT, 1977, p 136.

Warrant Officers Class I

The cap badge worn by warrant officers class I was a one-piece badge in gilding metal that was also copied from the RAF badge for the same rank. There were several variations of this badge. On the illustrated example, the bare metal between the crown and the eagle is painted black to give the appearance of a separate crown when seen against the black mohair cap band, and red melton material has been added to represent the cushion in the crown. All variations, with and without the black paint and red cushion, were worn during this period.

Field Service Cap

The field service cap was introduced in 1937 to provide an easily manufactured form of headdress for the thousands of recruits coming into the service. This cap had the advantage over the service dress cap in that it could be folded and put in the pocket—a great advantage for aircrew. The side flaps could be unbuttoned, pulled down over the ears and rebuttoned under the chin, giving some protection from cold winter winds.

A significant change from pre-1937 caps was that badges on field service caps were of the black oxidised type instead of the gilt badges of the service cap. Officers and Warrant officers wore the black oxidised eagle and crown badge and Flight Sergeants and below wore the Airman's cap badge on the left side of the field service cap. Oxidised cap badges were made by a variety of makers in both Australia and Canada and show small differences in detail. Some caps were also made in Britain. Overseas manufacture of badges and caps did not take place before World War II.

Wolseley Pattern Helmet

In the 1937 regulations, the description of the Wolseley pattern helmet was changed to 'drab helmet' and the dark blue fold in the puggaree had disappeared. The drab helmet was still worn with tropical dress and summer field service dress.

Drab helmet with chinstrap secured over the peak. This example does not have the correct RAF Flash which is maroon and dark blue with a thin blue central stripe. The description was not changed to RAAF Flash in the Dress Manual until about 1995

RANK INSIGNIA
OFFICERS, ROYAL AUSTRALIAN AIR FORCE, 1937

On the full-dress tunic, mess dress jacket and winter service dress tunic, rank was indicated by rows of gold braid, with a gilt eagle-and-crown badge worn ¼ inch above the top row of rank braid.

On the field service dress (winter and summer), rank was indicated by rows of black/blue braid on each epaulette, with an oxidised eagle and crown badge worn above the top row of rank braid.

Rank	Hat Badge	All Uniforms	Hat
Air Chief Marshal	Air Officer	1 row broad & 3 rows of ordinary braid.	Two rows of gold wattle leaves on patent leather peak.
Air Marshal	Air Officer	1 row broad & 2 rows of ordinary braid.	As above.
Air Vice-Marshal	Air Officer	1 row broad & 1 row of ordinary braid.	As above.
Air Commodore	Air Officer	1 row broad braid.	As above.
Group Captain	Officer	4 rows of ordinary braid	One row of gold wattle leaves on patent leather peak.
Wing Commander	Officer	3 rows of ordinary braid	Cloth peak without wattle leaves
Squadron Leader	Officer	1 row narrow braid between 2 rows of ordinary braid	As above.
Flight Lieutenant	Officer	2 rows of ordinary braid	As above.
Flying Officer	Officer	1 row of ordinary braid	As above.
Pilot Officer	Officer	1 row of narrow braid	As above.

Officers' Black/Blue Rank Lace

Broad: 2 inches wide

Ordinary: 9/16 inch wide

Narrow: ¼ inch wide

QUALIFICATION BADGES: AIRCREW

Pilot's Flying Badge

The gold bullion pilot brevet was worn with full dress and winter service dress. A miniature version of this brevet was worn with mess dress and mess undress. With tropical mess dress, the badge was mounted on a detachable brooch.

The 1937/1938 bullion pilot brevet worn by Wing Commander Richard 'Dick' Cresswell

With summer service dress and tropical dress, a brevet embroidered in white silk on a black background was fastened to the jacket or shirt by detachable clips or studs. The white embroidered brevet was also worn on the jacket of field service dress.

The pilot brevet worn with field service dress, summer service dress and tropical dress

Observer's Brevet

With the more advanced, long-range aircraft being brought into service, it was found necessary to re-introduce the role of Observer as a specialist member of aircrew. The first Observer course was completed in March 1937 and a flying badge, based on the RAF design of white on blue, was awarded to all graduates. It was also available in a drab version.

Many of the graduates from this course did so in the rank of corporal, which was followed by promotion to sergeant after twelve months of satisfactory performance in the role. This was confirmed by the experience of personnel from No 10 Squadron, who had gone to the United Kingdom to collect the new Sunderland aircraft and who subsequently remained there after the outbreak of World War II. The squadron history recorded that personnel encountered difficulties with the RAF Police who could not believe that such junior ranks were permitted to wear aircrew flying qualification badges.[21]

As with the pilot brevet, a gold bullion observer brevet was worn with full dress and winter service dress, with a miniature version worn with mess dress and mess undress. The white-on-blue brevet was worn on summer service dress, field service dress and tropical dress.

1937 gold bullion observer brevets *A 1937 white-on-blue observer brevet*

Air Gunner Badge

The air gunner badge, or 'winged bullet' badge continued to be worn by members trained in air gunnery. Usually these were members of the armament mustering and, as they flew only when required, they were not considered full-time aircrew. A black oxidised badge was introduced in 1937 to wear with field service dress,

as opposed to the gilt badge that was worn on the sleeve of service dress. The 'winged bullet' badge was replaced by an air gunner brevet in 1940.

---

[21] *Maritime is Number Ten*, K C Baff 1983 ISBN 095923960X.

QUALIFICATION BADGES: GROUND STAFF

Medical Officer's Badge

The 1937 Dress Regulations described the medical officer badge as 'winged caduceus of Mercury surmounted by crown'. In service dress, a gilt badge was worn but in field service dress, the badge was 'black'.[22] This badge was a copy of the one used by the RAF and was initially worn by medical officers only. Later, it was worn by all medical branch staff on both lapels of the service dress and field service dress jacket, or on the peaks of the shirt collar.

A medical officer's badge on a service dress jacket (Source: State Library South Australia) with gilt and oxidised ('black') medical collar badges

Dental Branch Badge

Although RAAF dental officers belonged to the Medical Branch until September 1940, they wore a dental officer badge. Like the medical officer badge, the dental branch badge was copied from the RAF and was worn in a similar fashion. A gilt badge was worn on service dress and the oxidised badge on field service and tropical dress.

The badge was described as 'a laurel wreath with the letters "DB" placed therein and flanked on either side by wings – all in gilding metal. In field service dress, to be black'.[23] The badge was worn by dental officers, dental mechanics and dental orderlies. The badge remained in service until January 1949 when it was replaced by medical insignia.[24]

Gilt Dental Branch collar badge   *Oxidised Dental Branch collar badge*

---

[22] While the Dress Regulations prescribe badges in black, this means what is generally known as oxidised.
[23] 1937 RAAF Dress Regulations
[24] Air Board Order N.2/49 dated 21 January 1949

Wireless Telegraphist Badge

The wireless telegraphist badge continued in use until 1941 recognising the skills of electronic communications members. It was worn by both wireless telegraph (W/T) operators and W/T mechanics. Some members of these musterings also flew as aircrew, performing the wireless operator role on the longer-range aircraft that were coming into service such as the Avro Anson. The 'winged thunderbolt' badge was introduced in the 1920s and is illustrated in Chapter 1.

OTHER BADGES

Shoulder Devices - Monogram Style

Officers holding honorary appointments such as personal aides-de-camp, honorary physician or honorary surgeon to the reigning monarch wore a gilt royal cypher of the monarch as a shoulder device on the epaulette of their jacket or shirt. Three sizes of the royal cypher were approved.

- The large cypher was worn with service dress, ceremonial dress and the greatcoat.
- The small cypher was worn with mess dress and mess undress.
- The miniature cypher was worn by officers who had previously held the appointment of Air Aide-de-Camp to the sovereign.

Royal cyphers of King George V and King George VI are illustrated. The wearing of the royal cypher insignia continues to the present day

Eagle Badge

An eagle badge embroidered in blue silk on a dark background, was specified for wear on the upper sleeve of the dark blue tunic and the greatcoat just below the shoulder for all airmen up to the rank of Warrant Officer class II. On khaki uniforms, an embroidered khaki badge was worn in the same place. The eagle badge was not worn by members of the Citizen Air Force.

Left: Third Pattern: Service Dress
Below left: Fourth Pattern: Service Dress
Below: Fourth Pattern: Summer Dress

Citizen Air Force

All badges detailed for the Permanent Air Force were also authorised for the Citizen Air Force (CAF) with the exception of the shoulder eagle. All CAF personnel continued to wear their distinctive tricolour CAF badge on both upper sleeves just below the shoulder. This was worn by all ranks in the CAF and was worn by airmen in lieu of the eagle badge. This badge was based on the one worn by the Australian Flying Corps in World War I. Members of the CAF served on a part time basis and the badge, with variations, continued in use until the CAF was re-organised into the RAAF Active Reserve in the 1980s.

Brassards

The following brassards, also called 'armlets', were introduced during this period in addition to those covered in chapter 1.

- Provost Marshal. This cloth brassard was 3½ inches wide and consisted of three bands—black, red, then black. Black letters 'P.M' were on centre band, surmounted by a gilt eagle and bullion crown. It was worn on the right arm. A Provost Marshal is a senior commissioned officer who commands the police in the relevant military service.
- Assistant Provost Marshal. This brassard was the same as the Provost Marshal brassard except that the letters 'A.P.M' were substituted for 'P.M'. An Assistant Provost Marshal is an officer subordinate to the Provost Marshal and acts as his deputy.

CHAPTER 3: THE RAAF GOES TO WAR

Australia's entry into World War II began with a radio broadcast to the nation by Prime Minister Robert G. Menzies on the evening of Sunday 3 September 1939, in which he said:

"Fellow Australians, it is my melancholy duty to inform you officially that, in consequence of the persistence by Germany in her invasion of Poland, Great Britain has declared war upon her and that, as a result, Australia is also at war."[25]

The change from peacetime to wartime brought huge changes to the RAAF. Over a period of five years, the RAAF expanded from a strength of 3,489 members on 3 September 1939 to 173,622 men and women at the end of the war in 1945.[26] All these members were trained, accommodated and clothed in Australia before serving in dozens of overseas locations in climatic conditions that ranged from the tropical jungles of south east Asia, the deserts of North Africa to near-Arctic conditions of northern Europe in winter.

Policies relating to uniforms also underwent huge changes. Instead of signing contracts for hundreds of uniform items, the RAAF was now demanding tens of thousands of items. To meet the demand, uniforms had to be made of easily obtained materials and quick and inexpensive to manufacture. In addition, RAAF members serving away from established RAAF stations often had to obtain uniform items locally, leading to variations in manufacturing style and materials.

In 1939, the Australian Government signed an agreement with United Kingdom, Canada and New Zealand to establish the Empire Air Training Scheme (EATS) the aim of which was to create an aircrew training system to train pilots, navigators, wireless operators and air gunners for service in RAF squadrons. The scheme required each air force to train to the same syllabuses and to standardise its personnel policies so that personnel could move from a training school in one country to an operational training unit or squadron in another.

Among the standardised personnel policies were those concerning uniforms. Although the RAAF retained its unique dark blue colour for its uniforms, badges and insignia were generally standardised with the ones in use by the RAF.

Policy Changes

During World War II, the following general policies relating to uniforms were in force.[27]

- Full dress, mess dress and mess undress uniforms were not worn.
- Field service dress was worn by all ranks on most occasions.
- Airmen wore the field service cap instead of the service dress cap. Officers could wear either.
- Orders, decorations and medals were not worn but ribbons were worn with field service dress.

The changes to each form of dress or uniform item are detailed in the following paragraphs.

---

[25] Australian War Memorial encyclopedia, www.awm.gov.au/articles/encyclopedia/prime_ministers/menzies
[26] Alan Stephens, *The Australian Centenary History of Defence, Vol II, The Royal Australian Air Force*, Oxford University Press, Melbourne, VIC, 2001
[27] Air Board Order N.230 dated 21 November 1939

SERVICE DRESS

During World War II, only officers wore service dress. Airmen wore the newly introduced field service dress or 'battle dress'. Under wartime regulations, the gilt buttons and belt buckle of the pre-war uniform were changed to black Bakelite or oxidised brass. The eagle and crown badges above the rank lace on the sleeves were no longer required, however, there is evidence that some officers continued wearing the eagle and crown during World War II. Where metal badges were worn, they were in black oxidised form instead of gilt.

Left: Officer's winter service dress tunic, bearing rank lace for Flight Lieutenant and a pilot brevet. Right: Summer service dress tunic made of khaki gaberdine with sergeant rank chevrons on the sleeve and an observer brevet. This uniform is shown with the khaki fur-felt hat which was commonly worn in tropical areas

Ceremonial dress

During the period of World War II, ceremonial dress was rarely worn. Dress for parades such as graduation was usually service dress, summer or winter as appropriate.

Tropical dress

Although khaki shorts and short sleeve shirt remained the official tropical dress, it was common for RAAF personnel to wear the khaki service dress long trousers and long sleeve shirt without the tie in tropical areas. Although the sleeves were commonly rolled up during the day for comfort, this uniform had the advantage of covering the arms and legs at night to guard against mosquitoes in areas such as Papua New Guinea.

Flight Lieutenant Les Osborne, an armament officer with No 71 Wing at Goodenough Island, PNG in November 1943

WORKING DRESS

War Service Dress and Battle Dress[28]

In the late 1930s, the British Army developed a new uniform that was more comfortable and utilitarian especially for mechanised infantry in the field. It was less restrictive to the wearer, used less material than existing uniforms and was warm even while wet. Being designed for battle conditions, the new uniform was khaki in colour and was known as 'battle dress'. Realising the need for a better working dress, in 1941, the RAF adopted the same style of uniform but in blue-grey colour. The RAF uniform was worn open at the neck with a shirt and tie while the Army jacket was buttoned at the neck. Initially, the uniform was issued to airmen but quickly came into use by officers, particularly aircrew who found it comfortable to wear in aircraft. By about 1942, the RAAF dark blue version of this uniform was introduced for use as a winter working dress, primarily for aircrew, and was officially called 'suit, aircrew' but, colloquially, the army term 'battle dress' was used. By 1943, the uniform was known as 'war service dress'.

War service dress was composed of a short, wool serge jacket (or blouse) that was buttoned up the front, although the buttons were covered by a fly front. The blouse was open at the neck and was worn with a blue-grey shirt and black tie. Two pockets on the front of the blouse were secured by buttons concealed by the pocket flaps. It was fastened at the waist by a belt which formed part of the waistband. The high-waisted wool serge trousers of the same colour were buttoned to the back of the blouse to prevent the blouse riding up. The sleeves had a forward curve built into them so that they were more comfortable to wear while performing aircraft maintenance, driving a vehicle or flying an aircraft. On the trousers, there was a large map pocket on the front near the left knee and a special pocket for a field dressing near the right front pocket (on the upper hip). Airmen's war service dress was made from wool serge fabric, often called 'Hairy Mary' but later, officer uniforms were often tailored from a finer-weave wool barathea.[29]

When first introduced about 1942, badging on the blouse was minimal. Non-commissioned officers wore their chevrons on the right arm only and officers had rank lace sewn on the epaulettes. No badging other

---

[28] The author acknowledges the contribution of Mr Chris Kanca (www.stby.com) in the preparation of this section.
[29] PG Hering, *Customs and Traditions of the Royal Air Force*, Gale & Polden Ltd, Aldershot, UK, 1961, pp 225–6.

than aircrew brevets were worn. However, by 1943, war service dress had become the 'everyday' uniform of the RAAF so dress regulations now required the same badges as were worn on service dress i.e. country titles, shoulder eagles, ribbons, rank lace and chevrons.

Until the RAAF dark blue version of war service dress became available in Europe in 1942, RAAF members serving with RAF units often wore the RAF blue-grey uniform, but with RAAF badges.

War service dress blouse with Flight Lieutenant rank lace on the epaulettes and the air gunner brevet above ribbons for Distinguished Flying Cross and the 1939/1943 Star. The blouse also has a Pathfinder gilt badge on the flap of left breast pocket and an 'AUSTRALIA' badge on each upper sleeve, indicating wear outside Australia

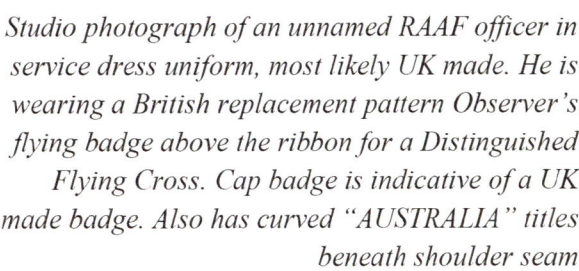

Studio photograph of an unnamed RAAF officer in service dress uniform, most likely UK made. He is wearing a British replacement pattern Observer's flying badge above the ribbon for a Distinguished Flying Cross. Cap badge is indicative of a UK made badge. Also has curved "AUSTRALIA" titles beneath shoulder seam

Personnel of No 460 Squadron, who were serving in the UK flying Wellington bomber aircraft. All are wearing war service dress, with field service caps and a range of aircrew flying badges.
Source: OAFH, Canberra

Khaki Battle Dress

It is also known that RAAF members serving in the Mediterranean region wore the khaki version of this uniform during the colder months or when flying.

Khaki battle jacket with warrant officer oxidised rank insignia and maker's label dated 1944

The RAAF crew of a No 454 Squadron Baltimore bomber, photographed in Italy circa 1944, are wearing a range of uniforms. The two on the right are wearing khaki war service blouse (British issue) with drab trousers. Source: OAFH, Canberra

Pacific Theatre

In the Pacific region, cold weather was unlikely to be encountered, so working dress was usually drab uniforms, either short or long sleeved.

An aircrew group that are probably a Catalina crew in northern Australia. All are wearing flying badges on their khaki drab shirts. The two officers (centre front row) have rank lace on their epaulettes; the warrant officer (front row, second from left) is wearing a wristlet on his right wrist displaying his metal rank badge and the NCO members are wearing chevrons on their right arms only. The shirts are a mix of long sleeve type (rolled up) and short sleeve issue. The two observers in the rear row (centre and far right) are wearing khaki drab flying badges. Source: RAAF Museum

TRAINEE AIRCREW DRESS

Initial Service Training

Under the Empire Air Training Scheme (EATS), large numbers of aircrew were recruited and trained in Commonwealth countries for service with RAF squadrons in Europe. Initially, four categories of aircrew were trained—pilots, observers (later renamed 'navigators'), wireless operators/air gunners and air gunners. Later in the war, additional categories of bomb aimers and flight engineers were also trained.

Australian aircrew trainees were trained in the following four phases:

- Ten weeks of initial service training, introducing the new recruits to life in the Air Force, was conducted at six initial training schools around Australia. Each trainee was assessed for their aptitude for each of the aircrew categories. At the end of initial training, trainees moved to bases and schools that would conduct the aircrew training they were selected for, i.e. pilot, air gunner, etc. Trainees wore the rank of aircraftman during initial training but advanced to the rank of leading aircraftman for the basic and advanced phases of their training.
- Most trainees commenced the basic phase of their aircrew training in Australia.
- For the advanced phase of their training, some trainees remained in Australia but many went to Canada or Rhodesia (now Zimbabwe). At the end of this phase, trainees who graduated were presented with their 'wings' and were promoted to Sergeant or Pilot Officer.
- Aircrew bound for Europe or the Middle East then moved to UK for training on an operational aircraft, followed by combat service in an RAF squadron. Aircrew destined for RAAF squadrons in the Pacific theatre completed their training on an operational aircraft in Australia before moving to an RAAF squadron.

Training in Canada

Many photographs of personnel leaving for Canada show them kitted in khaki drab uniforms which were appropriate for the voyage through tropical areas. Yet three weeks later they would be in the throes of a Canadian winter which was a complete contrast for Australians.

Aircrew trainees in summer service dress tunics leaving Australia for Canada in March 1941. The white flash on their field service cap identifies them as trainees.
Source: AWM, Canberra

The Canadian winters led to personnel being issued with a Canadian style piece of headgear– sometimes called a toque or often, after its manufacturer, a Brill cap. These caps were made in RAAF dark blue material but some examples of the caps in RAF blue-grey with RAAF badges also exist. An AMO directive stated that all issues of this style of cap to aircrew trainees were to be returned to stores prior to departure from Canada to Britain. Some made it back to Australia!

Brill cap made in RAAF Blue and fitted with a Scully, Montreal RAAF oxidised airman's cap badge. (Source David McDonald)

Brill cap made in RAF/RCAF blue-grey material and fitted with a Scully, Montreal RAAF oxidised airman's cap badge

The brill cap is being put to good use!

Newly graduated pilots in winter service dress at Fort McLeod, Alberta, Canada circa 1941. The day of graduation was the only day that they wore both their flying badge and the white trainee flash on their field service cap. The day after graduation, the flash was removed from the cap

Uniforms Post Aircrew Training

Graduates from the respective aircrew training courses were either commissioned with the rank of Pilot Officer (approximately 10 per cent of course members) or were promoted to the rank of sergeant. The newly commissioned officers were authorised to obtain tailored uniforms (2 sets of both summer and winter) and, depending on the initial posting and the timing of actual movement, the uniforms were mainly made in Canada or Britain. Pending receipt of the tailored uniforms, these newly commissioned officers continued to wear their existing airmen's uniforms, but with all airman rank, cap and shoulder insignia removed. In lieu, they wore a white brassard on their left arm to indicate that they were awaiting issue of new uniforms, which would be made by civilian tailors and could take four to six weeks to manufacture.

A newly commissioned officer wearing winter service dress with a white brassard indicating that he is awaiting the issue of new uniforms. All airman rank and badge insignia had been removed from this uniform (Source Australian War Memorial)

RANK INSIGNIA

Officers

Officers rank insignia continued to be lace stripes sewn on the lower sleeve on winter service dress uniforms. However, on summer service dress, tropical dress and war service dress, the rank lace was sewn on shoulder epaulettes or on rank slides that were worn on the epaulettes.

On summer service dress or tropical dress, the rank lace was more commonly sewn onto rank slides which were slid over a shoulder strap on the khaki tunic or shirt. The rank slide was then easily removed when the shirt/tunic was washed.

Rank lace indicating the rank of flight lieutenant sewn directly on the epaulette of a war service blouse

A pair of khaki rank slides with Flight Lieutenant rank and eagle and crown insignia. Note that the eagles are an opposing pair flying to the left and right

A khaki shoulder board for with Flight Lieutenant Rank lace sewn directly on the epaulette

Warrant Officer, SNCO and Airmen Insignia

The ranks of sergeant major class I and sergeant major class II, which were in use in 1921, were gradually replaced over the next two decades by warrant officer class I and warrant officer class II. In 1940, this was formalised when both sergeant major ranks were abolished and the one rank of warrant officer replaced the two classes of warrant officer.

The rank insignia worn by warrant officers remained the British coat of arms, which was the same as that worn by sergeant major class I and warrant officer class I. This badge was worn on both sleeves below the elbow. Warrant Officers class II wore a Tudor crown on both forearms.

When wearing a short sleeve shirt, or no shirt, in tropical areas, warrant officer or SNCO rank insignia was often worn on a wristlet of cloth, leather or metal.

Embroidered warrant officer badge for winter uniforms. There was also a variant with silver thread instead of light blue

Embroidered warrant officer badge for summer uniforms

Gilt and oxidised warrant officer badges worn on the summer or tropical dress shirt

Metal WO rank badges on cloth & metal wristbands for wear with short sleeve shirts

Crowns worn by Warrant Officers Class II and above chevrons by Flight Sergeants: Oxidised, Gilt & Embroidered

Embroidered flight sergeant rank insignia for winter uniform

Flight sergeant insignia for summer or tropical uniform with an oxidised metal crown

The crown in the flight sergeant badge could be embroidered or metallic on either summer or winter uniform. The chevrons could be three individual chevrons sewn onto a backing piece or they could be made as a single three chevron piece.

Sergeant rank insignia for winter and summer dress

Corporal rank insignia for winter dress

Corporal: summer and tropical dress

Leading aircraftman rank insignia for winter and summer dress

LAC oxidised copper rank insignia as worn during the war

A variety of uniforms and rank insignia as members of the Australian Spitfire Squadron say goodbye to their mascot "Sprog", who had to be left behind when the unit relocated in 1944

HEADDRESS

As all forms of RAAF headdress were manufactured in most states of Australia as well as Canada, the United Kingdom and India, considerable variation occurred in their appearance.

Service Dress Cap

During the war period, shortages of materials and the need to kit out thousands of recruits necessitated some changes to the manufacture of uniform items. One of these was a variant of the officer cap badge, generally known as the 'economy' badge. Instead of being hand sewn, the three components—crown, eagle and wreath—were made in a light gilt metal and fixed to a padded dark blue or black background with metal tabs.

The cap worn by officers with winter service dress with a wartime 'economy' cap badge

Officer's 'economy' cap badge

Maker's label on reverse: National Clothing Co (1939) Ltd, Perth

Service dress cap with khaki cover and the officer's 'economy' badge

Warrant officer service dress cap and badge

The service dress cap for flight sergeants and below was unchanged from the pre-war cap. It was rarely worn during the war years, as most airmen were only issued with the field service cap. Motor transport drivers were the only mustering that was authorised to wear the service dress cap during the war period.

Field Service Cap

As described in Chapter 2, the field service cap was introduced to provide an easily manufactured form of headdress for the thousands of recruits coming into the service. This cap had the advantage over the service dress cap in that it could be folded and put in the pocket—a great advantage for aircrew. The side flaps could be unbuttoned, pulled down over the wearer's ears and then buttoned again under the wearer's chin, giving the wearer some protection from cold winds in winter.

The badges on field service caps were of the black oxidised type instead of the gilt (gold) badges of the pre-war service dress cap.

Field service cap fitted with black oxidised eagle and crown badge worn by officers and warrant officers

Close-up of the K.G. Luke, Melbourne field service cap badge

In 1941, white peak flashes were introduced for the field service cap worn by trainee aircrew. The white flannel flashes were issued to recruits at the start of initial training and were discarded upon graduation. The flash was attached to the cap by a dressmaker's hook at each end of the flash.[30]

A field service cap with the black oxidised airmen's cap badge and white trainee peak flash

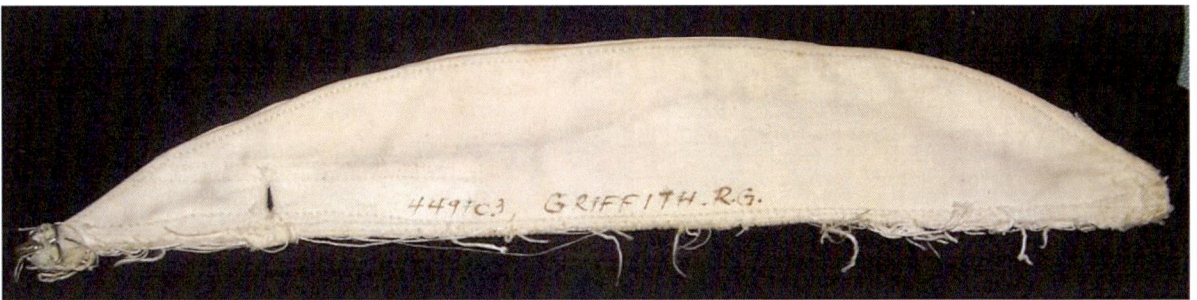

White flannel trainee aircrew peak flash with a dressmaker's hook at each end

---

[30] Air Board Order N.18 dated 3 January 1941

Khaki Fur-Felt Hat

When the materials for manufacturing the Wolseley Pattern helmet became scarce at the start of World War II, the khaki fur-felt hat became the standard form of headdress in tropical areas. Unlike a similar hat worn by the Australian Army, the RAAF fur-felt hat was never turned up on the left side, except when the wearer was marching with a Lee-Enfield .303 rifle fitted with bayonet.

The illustrated khaki fur-felt hat has a black oxidised airman's cap badge and the seven-fold RAAF puggaree with a blue stripe. A further illustration below shows a flat-style of puggaree with the oxidised airman's cap badge but without the traditional seven folds. This style was in use late in the war, probably to save on material and the cost of manufacture.

This flat style of puggaree was also used on the khaki fur-felt hat by officer's with the RAF flash as on the example below

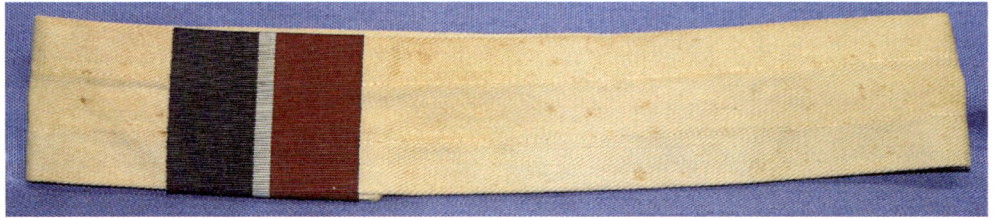

QUALIFICATION BADGES: AIRCREW

Badges and insignia were a highly valued part of the RAAF uniform and were an important part of unit and service identity. Serving in various parts of the world, RAAF members often had to obtain their uniform badges from local sources, which produced many badges with large variations from the authorised design and colour. Those shown below are just some of the badge variations which were worn by members of the RAAF.

Personnel with Prior Aircrew Qualifications

When the RAAF was formed in 1921, the service adopted a policy that allowed a member who had previously qualified for a flying badge to continue to wear that badge even when they were no longer employed on flying duties. During World War II, many who had served during World War I again put on a uniform to make their contribution. Although no longer able to fly operationally, previously qualified pilots and observers were authorised to wear the flying badges of their former services, including the Royal Flying Corps, the Royal Naval Air Service and the Australian Flying Corps.

RAAF Aircrew Flying Badges 1940

At the start of the Empire Air Training Scheme (EATS) in 1940, the RAAF trained four categories of aircrew—pilot, observer, wireless operator/air gunner and air gunner. The observer navigated the aircraft and released the bombs as well as operating cameras on reconnaissance missions. The wireless operator/air gunner had the primary duty of sending and receiving Morse code over the wireless (radio) but they were also trained to fire the guns when the aircraft was under attack.

Aircrew were identified by wearing one of three flying badges—pilot brevet (known officially as 'the flying badge'), the observer brevet or the air gunner brevet. Both the wireless operator/air gunner and the air gunner categories wore the air gunner brevet but wireless operator/air gunners also wore the wireless/telegraphist badge on their sleeve to indicate their additional skills.

The embroidered flying badges were usually white silk on a black background with the wreath in blue. Flying badges for wear on drab uniforms were light khaki silk on a darker khaki background. The observer badge, in particular, underwent many changes in design over the period of the war.

Pilot brevet, white on black background

Pilot brevet with khaki background

Early style Observer brevet *Later style Observer brevet*

Observer brevet on khaki background

Flying Badges 1942

In July 1942, the RAF reorganised their aircrew categories and created several new categories to take into account the new roles and technology that were rapidly changing military aviation. In December 1942, the RAAF changed their aircrew categories and flying badges to stay in step with the RAF. These changes were as follows:[31]

- The observer category was renamed 'navigator' to better reflect the role of this crewmember. The brevet was changed to show the letter 'N' instead of 'O'.
- The specialist air bomber category was introduced for the crewmember who operated the complex bomb sight equipment on heavy bombers.
- The flight engineer category was introduced for the technical crewmember who assisted the pilot in the operation of the engines and complex aircraft systems, such as the fuel system, in heavy bomber and large maritime patrol aircraft.

The new flying badges were in addition to those already in use.

Navigator

Air bomber

Flight Engineer

Air Gunner

---

[31] Ian D Jenkins, *The P.N.B.W. Aircrew Scheme in the RAAF*, Canberra 2008 – Privately Published

Flying Badges 1944

For the first four years of the Empire Air training Scheme (EATS), wireless operator/air gunners (WAGs) and air gunners wore the same 'AG' brevet. WAGs also wore the wireless/telegraphist badge ('hand grasping thunderbolts') on their sleeves.

However, most WAGs felt that wearing the AG brevet did not fully recognise the extensive training and assessment that they had undergone during their training. They would explain to visitors that WAGs took as long to train as pilots and much longer than air gunners. There was also a problem when WAGs were commissioned. By tradition, officers did not wear trade badges on their sleeves, so that when a WAG was commissioned, the only aircrew badge they could wear was the AG brevet.

In August 1944, the WAG category was renamed 'wireless operator (air)' and the signaller brevet, with the letter 'S', was introduced. The introduction of this brevet meant that the wireless/telegraphist sleeve badge was only worn by ground crew qualified in this trade. After the war, the wireless operator (air) category was renamed again 'signaller' category and its members continued to wear the S brevet.

The single-wing flying badges underwent a small change in 1944 when the wing became more 'outstretched'. This style continued in the single-wing flying badges used after the war.

Signaller (white on black)

Signaller (on khaki background)

Pathfinder Badge

The Pathfinder Force was an RAF force of squadrons manned by experienced aircrew. Their mission was to mark targets for the following bomber force. Members of this force were entitled to wear the Pathfinder badge on the flap of the left breast pocket. The badge could not be worn while flying on operations because of possible repercussions if the crewmember was captured by the enemy. Once qualified for the Pathfinder badge, the member wore this highly-prized badge for the rest of their career.

The first badges were the same as standard eagle-and-crown badges with two lugs but these were soon replaced with badges with brooch pins so that tell-tale holes were not left on the pocket flap. In the post war period, replacement badges were manufactured in Australia by Miller and Sons, Sydney for members of the RAAF who had been members of the Pathfinder Force during the war. The badges are distinguishable by the maker's mark on the reverse and the fact that they were constructed in two parts, with the top feathers and body of the eagle sweated onto a base of the wings and tail feathers.

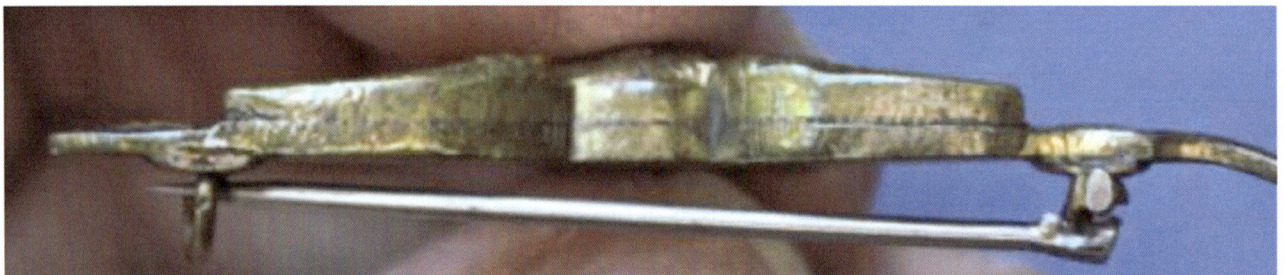

Pathfinder badge

Canadian Pattern Brevets

With a large proportion of EATS advanced aircrew training being conducted in Canada, the Royal Canadian Air Force (RCAF) arranged the manufacture of flying badges for presentation at graduation parades. Unfortunately, some Canadian manufacturers used the Canadian pattern for the badges but substituted 'RAAF' for 'RCAF'. The Canadian style badges differed from RAAF badges in the following two areas:

- The area inside the crown was embroidered in red on some badges.
- The single-wing badges were surmounted with a crown and had the letters 'RAAF' at the base of the wreath.

RAAF graduates had little choice but to accept and wear these incorrect badges. Considerable difficulty was experienced by Canadian-trained personnel when they arrived in Britain and they were told to replace the Canadian-made badges with approved ones as soon as practicable. However, this could take up to twelve months. Although not approved for wear on RAAF service dress tunics, some aircrew wore the Canadian pattern wings on their battle dress blouse, which were designated as aircrew working dress and, as such, could not be worn off base. Some examples of Canadian-made flying badges presented to RAAF graduates are shown below.

Canadian-made Pilot brevet

Canadian-made Navigator and Air Gunner brevets

Unauthorised Brevets

Due to the difficulty of obtaining uniform supplies in remote areas during wartime, many unauthorised variants of RAAF flying badges came into use.

The wireless operator/air gunner (WAG) brevet, in particular, was the source of many misconceptions. Only the RCAF had an authorised WAG brevet[32]; the RAAF and the RAF did not. As explained previously, wireless operators/air gunners wore the air gunner (AG) brevet on the left breast and the wireless/telegraphist badge ('hand grasping thunderbolts') on the sleeve until the signaller (S) brevet came into use in 1944. Any WAG brevet in the style of the RAAF single-wing brevets had been unofficially modified from another brevet and was unauthorised.

An example of an unauthorised, modified Wireless Operator/Air Gunner brevet. Source: AWM Canberra

---

[32] Imperial War Museum website, viewed 31 December 2017 at http://www.iwm.org.uk/collections/item/object/30069110

Another common error that occurred when brevets were made by unauthorised manufacturers was the use of an incorrect colour for the wreath. RAAF brevets were unique among the Commonwealth air forces in the use of blue for the wreath; other air forces followed the RAF in using brown for the wreath. However, some tailors, particularly in the Middle East and South East Asia, made RAAF brevets using RAF colours resulting in many brevets like the one shown below.

A Pilot brevet incorrectly made with the wreath in RAF colours.
(Source: RAAF Museum)

A Unique Brevet

James Rowland enlisted in the RAAF in 1942 and trained as a bomber pilot, serving in RAF Bomber Command and Pathfinder Force. In January 1945, he bailed out of his Lancaster bomber over Germany and became a prisoner of war, during which time his uniform was destroyed. To maintain his RAAF identity, and no doubt some sanity, Rowland hand-made this brevet from a small section of the lining of his flying boots and remnants of battle dress material, along with some thread. He brought this precious souvenir of his wartime service back to Australia after release from captivity. Rowland continued to serve in the RAAF after the war, becoming Chief of Air Staff in 1975. Air Marshal Sir James Rowland, AC, KBE, DFC, AFC, and also served as Governor of New South Wales from 1981 until 1989.[33]

---

[33] Alan Stephens and Jeff Isaacs, *High Flyers–Leaders of the Royal Australian Air Force*, AGPS, Canberra ACT 1996, p 158

This image illustrates the huge numbers of aircrew trained by the RAAF in World War II. Two civilian workers examining a production run of flying badges, somewhere in Australia circa 1942. Note that the four columns on the left are single wings which were made without the appropriate letter in the monogram, this being done later to meet the actual needs of the RAAF

QUALIFICATION BADGES: GROUND STAFF

Branch Identification Badges

See chapter 2 for examples of Medical and Dental Branch identifying badges. Although these two badges had been made in gilded metal or brass before the war, wartime policies required that, along with many other badges, they should be made of black moulded material, copper oxidised, or embroidered cloth.

Pharmacists 1941-1949

The first members of the RAAF Pharmaceutical Branch were civilian pharmacists appointed to the Citizen Air Force with the rank of Pilot Officer and whose names appear in the Air Force List of September 1940, with seniority dating from August 1940.

The Pharmacist Badge was probably introduced into RAAF service during the first half of 1941, appearing in oxidised brass or copper, with two lugs on the reverse for fastening to the collars by means of split pins. The design consists of the standard Medical Branch collar badge, or winged caduceus of Mercury, surmounted by a crown with the addition of a Gothic "P" superimposed on the serpents below the base of the wings. The badge was struck by Stokes of Melbourne.

The Pharmacist collar badge was worn in the same manner as the Medical Branch collar badge and is described in the RAAF Dress Regulations for August 1944 for wear as:

"Worn on the collar of the jacket, one at each side. The bottom of the badge is to be one inch above the inner end of the step opening and the staff of the badge to be parallel with the inside (rolled) edge and midway between the inside (rolled) edge and the outside edge of the collar. When Nos 6A, 6B or 6C is worn without either jacket or blouse war service, the badge is to be worn on the peak of the shirt collar"

Close up of the Pharmacist lapel badges on a tunic

The badge remained in service until abolished by Air Board Order N.2/49, dated 21 January 1949 and has not been replaced since. This badge is considered to be quite rare, with few known to have survived. Along with the Meteorological badge, it is unique to the RAAF.

Education 1940-1949

The RAAF Education badge is officially described as *"Crossed flambeaux surmounted by an eagle with outspread wings"* and is worn in pairs in the same position as Medical, Dental and Pharmacist badges. The badge was adopted directly from the RAF and was worn throughout World War II and the immediate post-war period in oxidised form. It was produced in two sizes – one for wear on the lapel of the service dress tunic, and a smaller size to wear on the collar of the khaki drab shirt. In 1948 a gilded brass badge was introduced for a short period. It was officially phased out in 1949, although it continued to be worn until sometime later.

Bomb Disposal 1942-1944

This badge was worn by officers, warrant officers and airmen qualified in bomb disposal. It was authorised on 25 May 1942 to facilitate the recognition of qualified personnel entering civilian areas containing unexploded bombs. It turned out to be a very short lived badge. The responsibility for disarming bombs in civilian areas was handed over to the Army and the badge was withdrawn from 24 March 1944. It was reintroduced after World War II with a different style wreath.

Bomb Disposal – Winter Dress *Bomb Disposal – Summer Dress*

Meteorology 1942

Like the Pharmacist badge, the Meteorology badge was also exclusive to the RAAF. It first appeared in Appendix 5/B/1 of 13 April 1942. Positioning is prescribed as: " The bottom of the badge is to be one inch above the inner end of the step opening edge and mid-way between the inside (rolled) edge and the outside edge of the collar. The arrow of the badge is to be in a horizontal position". It was worn by all personnel with a meteorology mustering. Meteorological Officers in the RAF were afforded the distinction of their own flying badge (subject to qualification) – a standard half wing with the letter M.

Wireless/Telegraphist Badge

The 'hand grasping thunderbolts' badge was worn by airmen and airwomen qualified in communicating by Morse code over the radio, including aircrew of the wireless operator/air gunner category. It was introduced early in World War II and was worn immediately below the eagle badge on both sleeves of the jacket. On the summer uniform, the badge was embroidered in brown silk on drab material. It replaced the 'winged thunderbolt' badge worn by wireless/telegraphists before the war. The badge was also manufactured in metal and fixed with lugs through backing plates.

Wireless/telegraphist for winter service dress *Wireless/telegraphist for summer dress*

Wireless/telegraphist gilt badge   *Wireless/telegraphist oxidised badge*

OTHER BADGES

Country-of-Origin Shoulder Badges

With large numbers of RAAF personnel serving overseas, particularly in RAF squadrons alongside servicemen from other Commonwealth countries, the Air Board saw the need for a badge to identify members as coming from Australia. In 1941, members serving overseas were directed to wear an 'Australia' badge on the top of their sleeves just below the shoulder. For airmen of flight sergeant rank and below, the country-of-origin badge was integrated into the eagle shoulder badge that they had always worn. For officers and warrant officers, the badge contained only the word 'Australia' with both ends curving downwards. As many of these badges were made by overseas manufacturers, there were many small variations in their design.

For winter service or battle dress, the badges were made with light blue silk on a dark blue background. For summer and tropical dress uniforms, the badges were brown silk on khaki material.

As these badges were only intended for personnel serving overseas, they were strictly controlled. Badges were issued to members immediately before embarkation for overseas destinations and were required to be removed immediately on return to Australia. In the case of airmen, they were to be replaced by an eagle shoulder badge.

Officer and warrant officer country-of-origin badge, for winter (left) and summer (right)

Airman's country-of-origin badge, for winter (left) and summer (right)

Operational Service Chevrons[34]

In March 1943, the RAAF introduced the practice of wearing chevrons to denote years of operational service. The term 'operational service' meant service in any theatre of operations which included Papua New Guinea and any part of Australia that was declared operational. Members were entitled to wear one chevron on the day they started their operational service, and an additional chevron on the completion of each twelve months' operational service. Personnel who were serving overseas on the 3 September 1939 were entitled to their initial chevron from that date.

Chevrons were of blue worsted embroidery, ¼ inch in width with the arms 1¼ inches long. They were worn on the right sleeve of the winter and summer service dress tunics, with the centre point upwards. Officers and warrant officers wore the chevrons above their rank insignia while airmen of flight sergeant and below wore them 4 inches above the sleeve cuff.

1943 chevrons, blue on blue *1943 chevrons, blue on khaki*
Both indicating between one and two years of operational service

Recognition of Former War Service 1943-1947

Many former members of the Australian Flying Corps, Royal Flying Corps and Royal Naval Air Service served again in World War II in various training, administrative and support capacities. From 1943, former war service was acknowledged by a single red chevron below the blue chevrons.

However, by 1944, the colours had reversed. World War II service was denoted by red chevrons while former World War I service was indicated by a single blue chevron above the red chevrons. All permanent RAAF, RAAFNS and WAAAF personnel were eligible to wear them.

When first announced on 5 January 1943, these chevrons were called 'overseas service chevrons'. However, when the order authorising the wearing of the chevrons was issued on 9 March 1943, the name had changed to 'operational service chevrons'. Later regulations in June 1944 referred to these chevrons as 'badges, war service'. These badges were withdrawn from use on 25 August 1947.

---

[34] Air Board Order A.64 dated 9 March 1943

1943 chevrons, blue on blue with red stripe, indicating three to four years of World War II operational service, plus service during World War I

1943 chevrons, blue on khaki, five stripes, indicating between four and five years of World War II operational service

1944 war service badges for operational service in World War II. Red on blue (left) represents four to five years and red on khaki (above) indicates one to two years

Wound Stripe 1944

In 1944, the Air Board approved the issue of wound stripes to members of the RAAF, RAAFNS and the WAAAF who were wounded due to enemy action. The stripes consisted of a woven stripe 1½ inches in length and 5/32 inches in width and was worn vertically on the left sleeve of the service dress tunic. Stripes denoting wounds sustained in World War II were gold and those for wounds sustained in World War I were red. More than one stripe could be worn and in the case of the person being entitled to wear both red and gold, the gold had precedence.[35]

---

[35] Air Board Order A.130 dated 1 May 1944

Brassards

With the greatly increased size and complexity of the RAAF during wartime, the need to rapidly identify essential personnel led to an expansion in the number of brassards worn by RAAF members. By 1945, the following brassards were in use.

- **Aerodrome Control Officer.**[36] Black cloth, 3¾ inches wide, with ¼ inch scarlet edge, and the letters 'D.P.' 2 inches high in scarlet cloth centrally placed.
- **Assistant Provost Marshal.** As pre-War. Illustrated in Chapter 2.
- **Provost Marshal**. As pre-War. Illustrated in Chapter 2.
- **Bomb Disposal Officer.** Light blue cloth, 3½ inches wide, with bomb separating letters 'B' and 'D' within laurel wreath embroidered in light blue on dark blue background. The brassard was worn while actually engaged on bomb disposal work.
- **Captain of Aircraft.** Blue cloth with a central vertical grey panel containing a red 'C' with a gilt eagle-and-crown badge above. The crew of any particular aircraft could be composed of personnel of a variety of ranks with, for example, a non-commissioned pilot and a commissioned navigator and other aircrew. Notwithstanding rank, the pilot was always in charge of the aircraft and so it was found to be necessary to have a Captain of Aircraft armband to reinforce the authority of the pilot over more senior personnel when flying. Many pilots, particularly in established crews, did not bother wearing this armband, but it was an official issue and no doubt an essential embellishment in many situations.

- **Medical Air Evacuations.** White cloth, 5 inches wide with central red Geneva Cross.

---

[36] Aerodrome Control Officer was previously called Duty Pilot and was in charge of aircraft movements within the designated aerodrome area – akin to a contemporary air traffic controller/joint battlefield airspace controller.

- **Medical Officer or Trained Nurse.** Scarlet/white/Scarlet horizontal bands of equal portions.
- **Newly Commissioned Officer.** All over white cloth, 4 inches wide, with a cloth tongue and four holes at one end and a single claw metal buckle at the other end. Worn by newly commissioned aircrew graduates in Australia and Canada following EATS training, whilst their tailor-made officer uniforms were being manufactured.
- **Duty Pilot.**[37] All over scarlet cloth as the pre-War brassard. (Chapter 1).
- **Intelligence Staff.** Light blue cloth, 3½ inches wide, with 1½ inch letters 'I.C.' (Intelligence Corps) embroidered in green silk centrally placed.
- **Movement Control Officer.** Scarlet cloth 3¾ inches wide, with letters 'M.C.' ¾ inch in height embroidered in black silk surmounted by eagle and Tudor crown in gilding metal.
- **Motorcycle Despatch Rider.** White cloth upper band with blue cloth lower band of equal portions. This armlet was required to be worn on BOTH arms by the despatch riders.

- **Red Cross (Geneva Convention).** Worn by personnel on medical duties including medical orderlies and aircrew serving with air ambulance units.
- **Non-Commissioned Officer of the Watch and Orderly Non-Commissioned Officer.** Black cloth, 3¾ inches wide, with ¼ inch scarlet edge bands. Another version had the scarlet letters 'O.S.' for orderly sergeant. Compare the red cloth and the dark blue stripe on blue pre-War brassards used for these two duties (Chapter1).
- **Orderly Officer.** Black background with scarlet edge bands and letters 'O.O.'. Compare the dark blue stripe on blue pre-War brassard. (Chapter1).
- **Orderly Officer WAAAF.** White paint stencilled letters 'WAAAF' on blue uniform material.
- **Railway Transport Officer.** Scarlet cloth 3¾ inches wide, with letters 'R.T.O.' 2 inches in height in black cloth centrally placed.
- **Railway Transport Staff.** Scarlet cloth 3¾ inches wide, with letters 'R.T.S.' in black.
- **Security Control of Air Travellers.** White cloth background with red lettering 'Security Control'.
- **Service Police.** Black background with white edges and central white letters 'SP', pictured below. Compare the pre-War brassard with scarlet letters 'RAAF SP' on black or dark blue uniform material.(Chapter 1).
- **Stretcher Bearer.** Cloth with letters 'SB'.
- **Welfare Workers.** Deep cherry cloth, 3½ inches wide, with designating letters ¾ inch high surmounted by badge of organisation. Worn by members of charitable organisations such as the Salvation Army.

---

[37] See footnote 13 (The Australian Centenary History of Defence, Vol II, The Royal Australian Air Force)

Authorised Uniform Manufacturers

During the course of the war, many civilian tailors in Australia were authorised to manufacture uniforms complete with insignia. Each authorised manufacturer was allowed to purchase material and supplies for these uniforms from RAAF stores. These manufacturers were each given an authority number, such as 'N.43', as a means of identifying their status as a manufacturer. Each authority number commenced with a letter indicating the state location.

- N   New South Wales
- Q   Queensland
- S   South Australia
- V   Victoria
- W   Western Australia

In other cases, the manufacturer was clearly identified by their business name. Melbourne Textiles (M.TX) is a common label and was a primary supplier during World War II. Post-war it became the Commonwealth Government Clothing Factory.

Some overseas manufacturers were also authorised to buy materials and tailored RAAF uniforms, for example, Gieves Ltd of London.

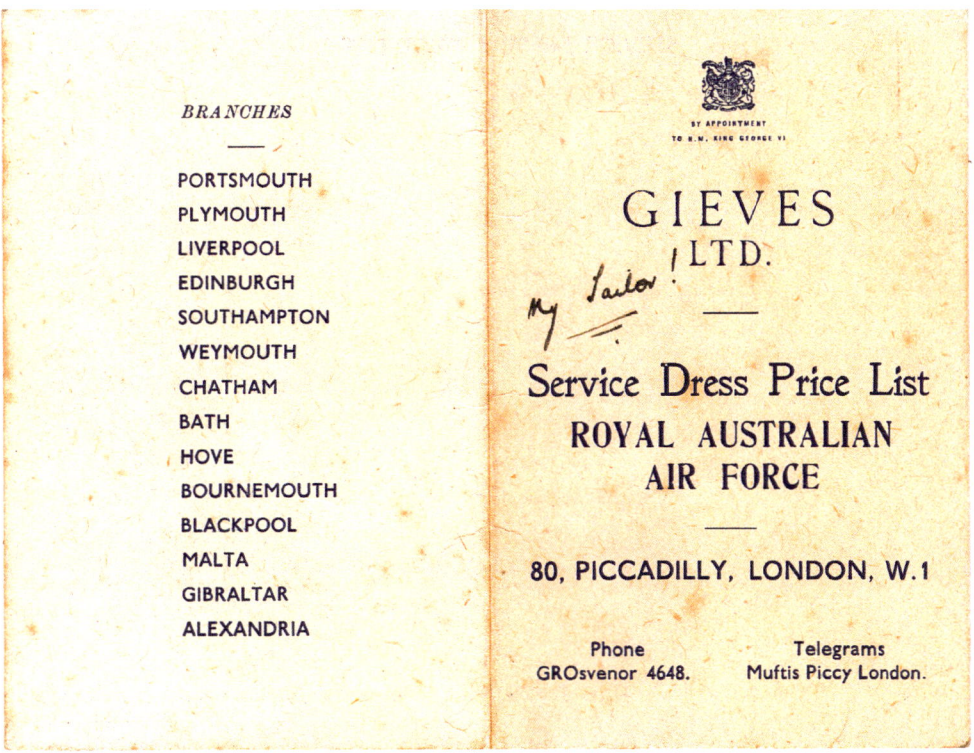

Service Dress Price List – April 1942 Gieves Ltd (now Gieves and Hawkes Military, 1 Savile Row, London)

SERVICE DRESS PRICE LIST

Royal Australian Air Force

	£	s.	d.			£	s.	d.
1 Blue Regulation Greatcoat *(including P/O Shoulder Straps)*	11	11	0	Braces ... from		5	0
Purchase Tax extra, nett ...		1	5	6	Cardigan or Pullover ...	1	15	9
1 Regulation Tunic ... *(including P/O Braid) Eagles extra*	8	0	0	Slipover (no sleeves) ...	1	5	0
Purchase Tax extra, nett ...		17	9	Wool Scarves ... from		8	6
1 pair Regulation Slacks ...		3	3	0	Handkerchiefs dozen ,,		15	0
Purchase Tax extra, nett ...		7	0	Pyjamas ... ,,	1	1	0
1 Regulation Raincoat ...		5	5	0	Small Bath Towels each		4	6
1 Uniform Cap and Badge ...		2	5	6	Large Bath Towels ,,		9	6
1 Field Service Cap and Badge		1	9	6	Winter Underwear various			
1 pair Regulation Shoes from	1	16	6	Summer ,, per suit		14	0
2 Blue Service Shirts with 2 collars each 17/6	1	15	0	Canvas Gym Shoes pair		8	6
1 pair Tan Regulation Gloves		13	6	**BADGES (Tax extra).**			
3 pairs Black Uniform Socks from @ 3/9		11	3	Pilots Wings per pair		5	6
1 Black Uniform Tie from		3	9	Observers Wings ,,		4	6
	£39	4	3	Air Gunners Wings ,,		4	6
Less discount for cash	1	16	9	Medical Badges ,,		7	6
	£37	7	6	Dental Badges ,,		8	6
				Clerical Badges ,,		15	6
				Sleeve Eagles ,,		6	6

Owing to market fluctuations all prices shewn in this list are subject to alterations without notice.

APRIL 1942

Gieves Ltd Service Dress Price List – April 1942 - continued

Sergeant Jack Rae selling RAAF uniform items at a Mobile Uniform Shop somewhere in Europe. The officer being served is wearing a khaki war service dress blouse (British issue) with what appear to be RAAF khaki drab trousers and a dark blue service dress cap.

Personnel of 460 Squadron, who were serving in Bomber Command flying Wellington aircraft. All are wearing war service dress and a range of aircrew flying badges.
Source: OAFH, Canberra

Officers of No. 71 Wing shortly after arrival at Goodenough Island, Papua New Guinea, November 1943. All are wearing khaki drab uniforms. Source: L Osborne Collection

CHAPTER 4:
WOMEN'S SERVICES 1940–1977

The women's services, which complemented the all-male Royal Australian Air Force (RAAF) for many decades, started in World War II but remained as separate services until the women's services were integrated into the RAAF in 1977.[38] The RAAF Nursing Service (RAAFNS) existed from 1940 until 1977, but the Women's Auxiliary Australian Air Force (WAAAF) was a temporary war-time service, approved in February 1941 and disbanded in 1947. The ensuing Women's Royal Australian Air Force (WRAAF), however, was a permanent women's service which was formed in 1951 and worked alongside the RAAF until integration in 1977.

The uniforms worn by the women of these services are covered in this chapter. The ones worn by female members of the RAAF after 1977 are covered in chapters six and seven.

ROYAL AUSTRALIAN AIR FORCE NURSING SERVICE

Until 1940, male nursing orderlies or civilian nurses attended to sick RAAF members. The civilian nurses at Air Force bases provided good care but, with the expansion of the service and the deployment of squadrons overseas, it became clear that a nursing service was required within the RAAF. Professionally trained nurses could also be used to train and supervise nursing orderlies in RAAF hospitals, allowing a rapid increase in the numbers of nursing staff.[39]

Group Captain (later Air Vice-Marshal) Edward Daley, Deputy Director of Medical Services, returning from an exchange posting with the Royal Air Force (RAF) in 1940 had been impressed with the organisation of the RAF Medical Directorate and in particular the Princess Mary's Royal Air Force Nursing Service. Modelled on this service, the RAAFNS came into existence on 26 July 1940.[40] Between 1940 and 1955, over 600 nursing sisters joined the service, serving in World War II, the Korean War and the Malayan Emergency[41] RAAFNS sisters also served in the Vietnam War.

Uniforms

"The principal problem concerning me in this was the design of an appropriate uniform. The tendency of my equipment people was to use as many as possible of the garments they had in store, which of course were for men, and I remember their trying to put an airman's khaki hat into the nurses' best kit. I believed that nurses would not be attracted to the Service unless we had a smart uniform for them and I think we produced one. I believe a woman has every right to be able to make the best of her appearance; anyway we attracted good nurses, 492 of them".[42]

The initial uniform was similar to that of Princess Mary's Royal Air Force Nursing Service.

---

[38] Alan Stephens, The Australian Centenary History of Defence, Vol II, The Royal Australian Air Force, Oxford University Press, Melbourne VIC, 2001, p 207

[39] Walker, Allan S, *Medical Services of the RAN and RAAF*, Australian War Memorial, Canberra ACT, 1961

[40] Gay Halstead, *The Story of the RAAF Nursing Service,* Nungurner Press, Metung VIC, 1994

[41] Gay Halstead, *The Story of the RAAF Nursing Service,* Nungurner Press, Metung VIC, 1994, preface

[42] *These are Facts* by Sir Richard Williams ISBN 0642993998

Service Dress

In the RAAFNS, service dress was also known as 'walking out dress' or 'outdoor dress' in both winter and summer forms. This was the dress that was worn in public or on parade.

Winter Outdoor Uniform

The winter outdoor or walking out uniform consisted of a dark blue Norfolk jacket and straight skirt in whipcord or barathea material, worn with a white blouse and black tie. Fawn stockings, black lace-up flat shoes and a dark blue felt hat completed the uniform. A woollen greatcoat of dark blue or a belted dark blue raincoat were worn in inclement weather.

RAAFNS walking out uniform with nurse's cape (or tippet) with senior sister rank braid and four red chevrons on the sleeve indicating four years of overseas service

Summer Outdoor Uniform

The summer outdoor uniform was similar to the winter one except the jacket and skirt were in khaki gabardine material and the hat was in khaki felt. Later, the white blouse was replaced by a khaki one.

The RAAFNS greatcoat was similar to the male version except it buttoned to right. Source: RAAF Museum Point Cook

Khaki jacket and skirt of the summer walking out uniform. Note that rank insignia for the summer uniform were on epaulettes

Three RAAFNS officers in summer walking out uniform in 1947. Source: RAAF News, May 1970

Korean War and the 1950s

Walking out dress jacket with sister rank braid in the 1950s

Vietnam War and the 1960s

By the 1960s, the walking out dress jacket had been redesigned without the belt. The buttons and badges were now gilt.

The uniform of an RAAFNS group officer in the 1960s

Four RAAFNS sisters in (left to right) ward dress, winter walking out dress, summer walking out dress and khaki aeromedical evacuation dress in the late 1960s

1970s

Changes were announced in late 1974, with news that the short cape was to be replaced by a hip length cape of light blue material with red lining. The new uniform encompassed elements of the all-seasons uniform with which it was designed to be worn, with or without the jacket. With the new style of jacket, rank was worn on the epaulettes instead of the sleeve.

RAAFNS jacket in the all-seasons uniform from the mid-1970s

The summer frock-style dress was worn by both RAAFNS and WRAAF members in the 1970s

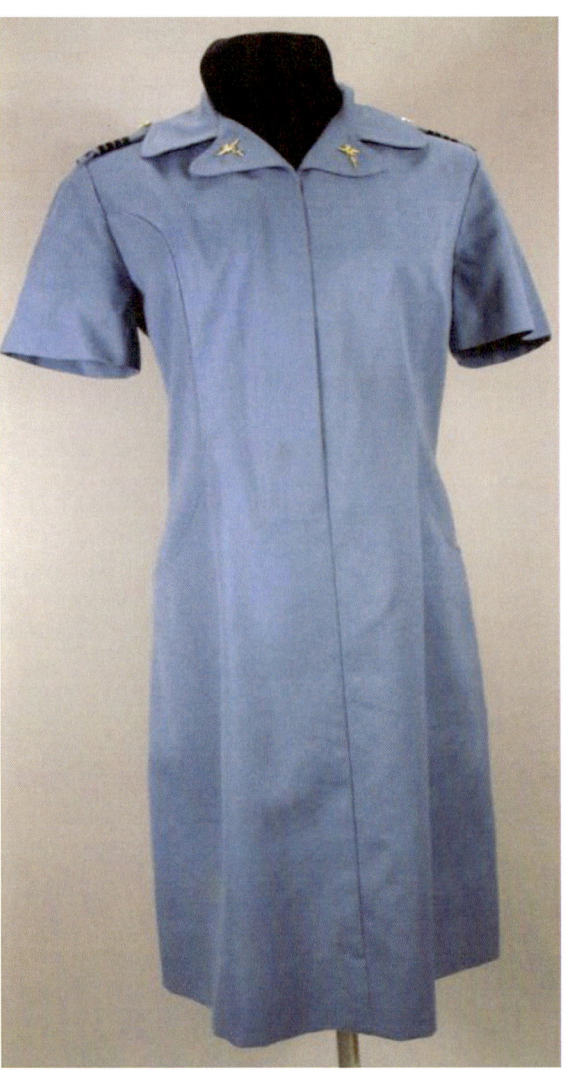

Ceremonial Dress

Although ceremonial dress was not worn during the war years, after the war, gilt buttons and badges reappeared on RAAFNS uniforms. Ceremonial dress was generally walking out dress with medals.

Mess Dress

The mess dress for RAAFNS sisters introduced after World War II was a simple belted dress of blue silk with long sleeves, detachable white cuffs and Peter Pan collar, with gilt Air Force buttons. Gold rank lace was worn on shoulder straps and black court shoes and dark brown silk stockings were worn. A white organdie veil completed the ensemble.

This mess dress was replaced about 1965 with a version which was also worn by WRAAF officers. The new mess dress was a dark blue synthetic dress with white collar, matching belt and anodised buttons. Rank insignia were worn on shoulder boards.

Left & above: Ward dress and cape worn during the 1940s

Far left: RAAFNS sister's mess dress worn in the 1950s. Source: RAAF Museum

Ward Dress

When RAAFNS sisters worked in RAAF hospitals and medical sections on bases, their working dress was a thick white cotton button-down uniform with black Air Force buttons and rank lace sewn on shoulder straps. A dark blue cape was worn over the ward dress when going outside in cold weather. White stockings, white shoes and a white organdie veil completed the uniform.

By 1974, the ward dress had been redesigned, the veil had shrunk to a white organdie cap and a blue-grey cape for wear over the ward dress had been introduced.

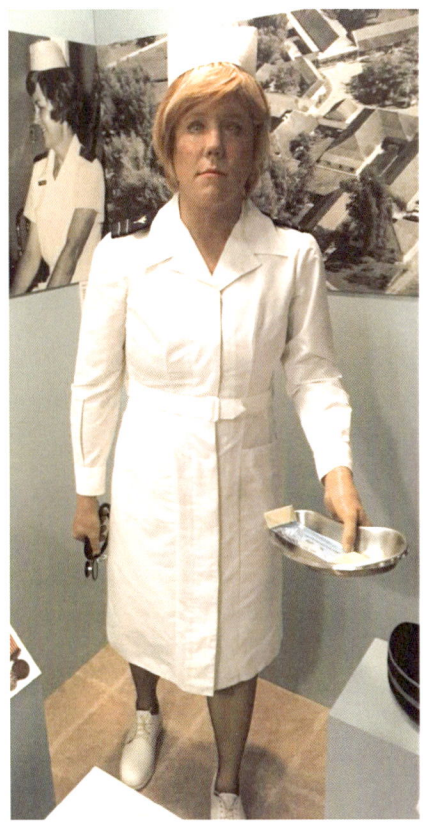

Ward dress (right) and blue-grey cape with red lining (left) of the mid-1970s

Tropical Working Dress

With the deployment of nursing sisters to the forward areas of Darwin and the South Pacific during World War II, the white ward dress and veil made these members highly visible from the air and were impractical in the field environment. So, in early 1942, instructions were sent to Darwin for uniforms to be dyed drab using strong tea. With the high rates of malaria in these areas, another uniform for wear between sunset and sunrise was introduced. This consisted of drab long-sleeved shirts with shoulder straps for rank insignia, drab slacks, black boots and canvas gaiters. These were commonly worn with fur-felt hats with RAAF puggaree and RAF flash.

RAAFNS sisters wearing tropical working dress and fur-felt hats at Madang PNG circa1945.
Source: AWM Canberra

Aeromedical Evacuation Dress

An increasingly important role of RAAFNS sisters during World War II was caring for wounded personnel while they were flown from combat areas to locations where permanent hospital facilities were available. Many of these aeromedical evacuations were flown from airfields in Papua New Guinea and islands north of Australia to Australian cities. Various uniforms were trialled for staff on these missions but the most suitable was the khaki tropical working dress, except that the dark blue field service cap was found to be more practical inside the aircraft than the fur-felt hat. For high-altitude, long-range flights when the cold temperatures became a problem, a blue fur-lined flying jacket was worn.[43]

---

[43] Allan S Walker, *Medical Services of the RAN and RAAF*, Australian War Memorial, Canberra ACT, 1961, p 414

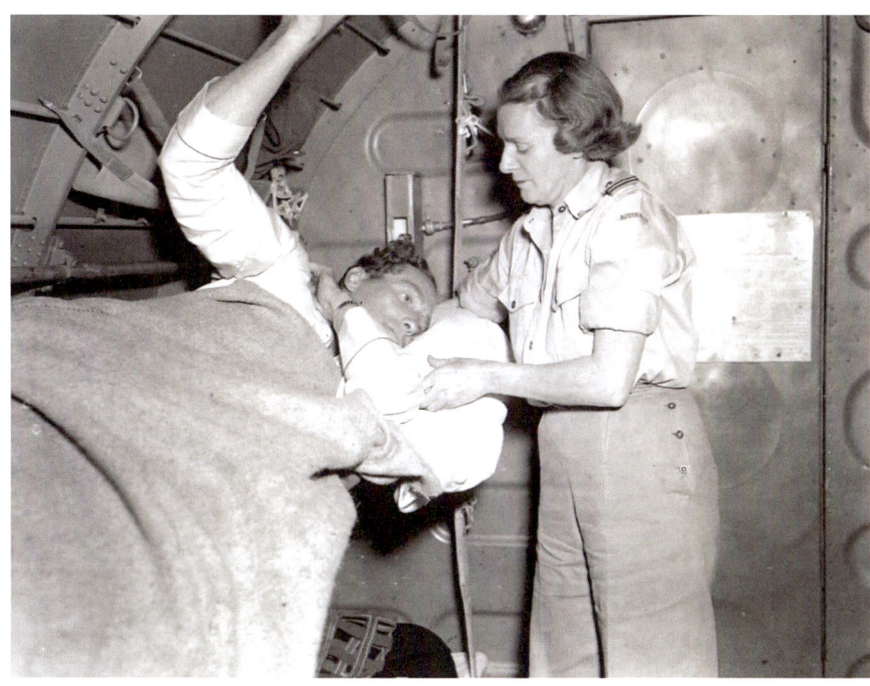

An RAAFNS aeromedical evacuation nurse dressed in khaki slacks and shirt attends a patient during the Korean War
Source: AWM Canberra

The khaki slacks and shirt were satisfactory in a tropical climate but, for winter in Japan and Korea, the RAAF dark blue battle dress was modified to provide a warmer but practical uniform for RAAFNS sisters on aeromedical evacuation missions. The battle dress was worn with black boots and the dark blue field service cap.

War service uniform (or battle dress) as modified for use by RAAFNS personnel on aeromedical evacuation duties. Note the slanting of the breast pockets to accommodate the female body shape. Additionally, the trousers are fitted with side buttons on the right hand side. The badge on the upper left sleeve is that of British Commonwealth Forces that were part of the Occupation Force in Japan after World War II

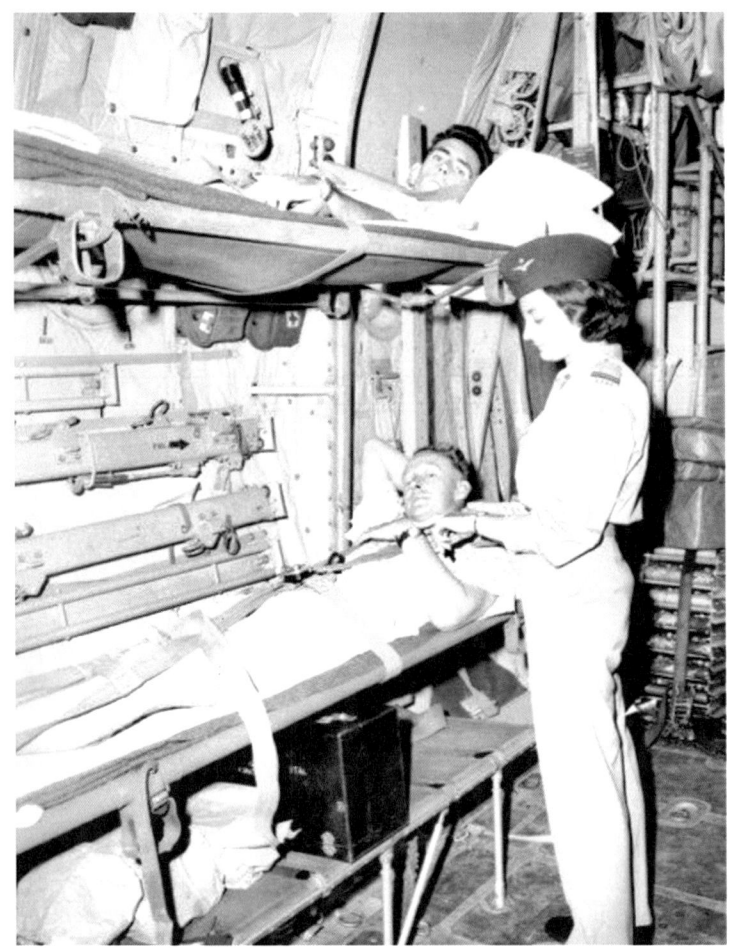

For aeromedical evacuation flights during the Vietnam War in the 1960s, the RAAFNS uniform was khaki slacks with matching over blouse, manufactured in drab material and worn with the dark blue service cap. Weatherproof flight line jackets were worn over this uniform whenever required by weather conditions.

Source: AWM Canberra

RAAFNS Rank Insignia

Because of their qualifications and professional status, all members of the RAAFNS held a rank equivalent to RAAF officers. When the service was formed in 1940, the 5-tier rank structure was based on that used in civilian hospitals. In 1955, the names of the ranks changed to the equivalent ranks in the WRAAF. In 1977, when the RAAFNS integrated with the RAAF, the RAAFNS ranks changed to the equivalent RAAF rank.

1940	1955	1977
Matron-in-Chief	Group Officer	Group Captain
Principal Matron	Wing Officer	Wing Commander
Matron	Squadron Officer	Squadron Leader
Senior Sister	Flight Officer	Flight Lieutenant
Sister	Section Officer	Flying Officer

Insignia for RAAFNS ranks of (left to right) principal matron, matron, senior sister and sister as worn on khaki shirts and jackets

RAAFNS officers wore their rank insignia on the sleeve of their winter jacket but on the epaulettes of the summer khaki jacket, the formal mess dress and all forms of working dress.

Headdress

With their walking out dress, RAAFNS officers wore a dark blue hat with a black woven band and an officer cap badge. After the war, when several RAAFNS officers reached the rank of group officer, they wore a black hat band with gold stripes on the top and bottom.

RAAFNS service dress hat with officer's economy cap badge from World War II era

RAAFNS service dress hat with officer's cap badge from the 1948–1954 period

RAAFNS Group officer's hat band with officer's cap badge of 1954–1970 period

The walking out hat worn by RAAFNS officers in the mid-1970s

Field service cap often worn by RAAFNS sisters on aeromedical evacuation flights in the 1960s

The blue-grey all-seasons female working dress cap was worn by nurses from the early 1970s

Medical Branch Badges

The medical branch badge containing the caduceus of Mercury (as used by the RAF) was worn on all uniforms and capes by RAAFNS officers. In keeping with wartime policies, these were issued in black oxidised colour and, after the war, they were replaced with gilt badges. With the accession of Queen Elizabeth II in 1953, the crown on the badge was changed to the St. Edwards' crown. (See chapter five for more details on the changeover to the St Edward's crown).

As explained in chapter five, the medical branch badge was changed to the Rod of Aesculapius in the late 1960s.

Medical branch badge based on the Caduceus of Mercury (left) with Tudor crown (to 1954) and the later badge introduced in the 1960s based on the rod of Aesculapius (right)

Integration

On 1 May 1977 the RAAF Nursing Service was integrated into the RAAF in keeping with the Australian Government policy of equal opportunity for females. At the changeover, the service numbered about 100 personnel. With this change, the Medical Branch created a nursing officer category which was open to male and female officers.

WOMEN'S AUXILIARY AUSTRALIAN AIR FORCE

The Women's Auxiliary Australian Air Force (WAAAF) was formed in March 1941 after considerable lobbying by women keen to serve and supported by the Chief of Air Staff who wanted to release male personnel serving in Australia for service overseas. The WAAAF was the first and largest of the World War II Australian women's services.

A WAAAF training depot was established at Malvern, a suburb of Melbourne, and the first 19 airwomen commenced training on 17 March 1941. Ten of the first enlistees were teleprinter operator trainees who were urgently needed to meet a deficiency of male wireless telegraphists. Airwomen were accepted into 73 different musterings (trades), including highly skilled positions in aircraft maintenance. Airwomen filled positions in many clerical, medical, transport, catering, equipment, meteorology, intelligence, signals and radar fields of employment.[44]

Group Officer Claire Stevenson, Director WAAAF, in her service dress uniform.
Source: AWM Canberra

By the end of 1941, some 1500 airwomen were serving. This number grew to a peak strength of 18,667 officers and airwomen by October 1944. They served in all states of Australia but were restricted to the area south of the line from Cairns in Northern Queensland to Geraldton in Western Australia. They were also prohibited from serving overseas.

Service dress

The WAAAF service dress uniform was similar to the equivalent male uniform except that a skirt replaced the trousers and the tunic was tailored to the female body shape. The tunic was a belted, button-fastened tunic with four flapped and buttoned pockets. WAAAF-issued tunics always buttoned-up in the male fashion (left-over-right) rather than in the usual female clothing style of right-over-left. This is because campaign and decoration ribbons are always worn on the left breast and could be partly obscured by the usual female-style fastening.[45] Despite this policy, some tailor-made WAAAF uniforms followed the female dress convention and buttoned right-over-left as in the illustrated example. The tunic was worn over a blue-grey shirt with a black woollen tie.

---

[44] Anne Heywood, *Women's Auxiliary Australian Air force (WAAAF) 1941-1947,* The Australian Women's Register website, 3 December 2002, viewed at www.womenaustralia.info/biogs/AWE0400b.htm 26 May 2015

[45] *WAAF Service Dress Uniform Buyer's Guide,* dated 16 November 2015, viewed at The Wadham's family website,

WAAAF officer uniforms were made of a good quality dark blue wool barathea material whereas the non-commissioned uniforms were made of a slightly coarser woollen cloth, though not as coarse as the 'Hairy Mary' serge material of the male uniform. The skirt was a plain, straight, two-gore style that finished between 14 and 16 inches above the ground. WAAAF skirts always ended at or slightly below knee height; they were never worn above the knee. Slacks were an alternative.

On the winter tunic, officers wore their rank on the lower sleeves while non-commissioned members wore it on the upper sleeve. This uniform was worn with blue socks and black shoes although silk or rayon stockings were commonly worn instead of socks.

A WAAAF officer service dress winter uniform

Squadron Officer Mabel Miller in khaki summer service dress uniform with tunic. Source: AWM Canberra

The summer service dress consisted of a tunic and skirt made of khaki gabardine material with gilt buttons and belt buckle. It was worn over a khaki shirt and black tie. Officer rank braid was on shoulder boards instead of the sleeve.

An alternative form of summer service dress was the khaki short-sleeve shirt and skirt. This was usually worn with the blue service dress cap, although a fur-felt hat could be worn in tropical areas. The uniform could be worn with or without the black tie. Officer rank slides were worn on the epaulettes while non-commissioned members wore their rank on the sleeve. Khaki shorts or slacks could replace the skirt when worn as working dress.

Summer service dress uniform of khaki short-sleeve shirt worn with skirt (above) or shorts (right)

Dress worn by WAAAF (or possibly WRAAF) nursing orderlies in RAAF hospitals and medical sections

Tropical dress

WAAAF members rarely served in tropical areas. Those members who did, usually wore overalls or the khaki shirt with skirt or slacks and a khaki fur-felt hat.

Working dress

WAAAF members served in many different trades and wore a variety of forms of working dress, depending on where they worked. Dark blue overalls with a dark blue wool beret were common working dress when performing aircraft maintenance and other technical jobs. The overalls were usually worn over the blue-grey service dress shirt and black tie. Personnel employed in clerical, intelligence and communications areas usually wore an appropriate form of service dress.

WAAAF members wearing overalls over their blue shirts and black ties as well as dark blue berets.
Source: State Library of South Australia

Sergeants Hilary Benn (right) and Glen Morton wearing overalls and service dress caps while testing a high-speed aircraft camera in the cockpit of an RAAF Avro Anson

A sergeant cook in her cook's uniform

When in the field, a common WAAAF working dress was khaki shirt and slacks worn with boots, and, if necessary, gaiters. Under cold conditions, a battle-dress blouse and field service cap were practical as shown above

Rank Structure and Insignia

The rank structure for the WAAAF was as shown below, with equivalent RAAF ranks shown on the right. The rank insignia was the same as that worn by the equivalent RAAF member. See Chapter 3 for illustrations of the rank insignia.

WAAAF	RAAF
Commandant-in-Chief	Air Marshal
Air Chief Commandant	Air Vice-Marshal
Air Commandant	Air Commodore
Group Officer	Group Captain
Wing Officer	Wing Commander
Squadron Officer	Squadron Leader
Flight Officer	Flight Lieutenant
Section Officer	Flying Officer
Assistant Section Officer	Pilot Officer
Under Officer	Warrant Officer
Flight Sergeant	Flight Sergeant
Sergeant	Sergeant
Corporal	Corporal
Leading Aircraftwoman	Leading Aircraftman
Aircraftwoman	Aircraftman

Headdress

WAAAF service dress cap with bullion officer badge on RAAF blue material background. Assistant Section Officer to Wing Officer

The service dress cap worn by Group Officer Claire Stevenson

The WAAAF officer cap was made of blue-grey wool barathea with a black mohair band and a patent leather chinstrap. The crown of WAAAF caps was softer and more convoluted than that on the male caps. Officers wore the same gold bullion cap badge on RAAF blue material as RAAF officers. Other ranks wore a cap badge of the same design as an airman's badge but which was embroidered in light blue silk on a dark blue background. A khaki fur-felt hat could be worn with summer uniform instead of the service dress cap and this was fitted with an airman's oxidised cap badge. A distinguishing feature of all WAAAF caps is that the chinstrap was held by two hooks rather than the usual buttons.

World War II WAAAF driver's cap with leather peak and oxidised metal cap badge. This cap is named to 110738 Thelma May Rule, who enlisted at Melbourne on 13 August, 1943, and was discharged as an aircraftswoman from Laverton Transit Departure and Reception Centre on 8 January, 1947

Airwoman's service dress cap with the embroidered cap badge

WAAAF fur-felt hat with the oxidised airman's cap badge

Qualification Badges

Members of the WAAAF generally wore the same qualification badges as members of the RAAF. See the qualification badges section of chapter 3 for images and more details.

The only badge specific to the WAAAF was the white enamel 'SP' badge worn on the lapels of the tunic when performing service police duties. These badges continued to be worn by service police members of the WRAAF until the early 1970s.

WAAAF/WRAAF Service Police Lapel Badge

Accoutrements

Dark brown leather gloves were often worn with service dress.

When Peace Returned in 1945

Once peace returned in August 1945, the WAAAF, being a temporary service, was no longer needed. Having made a major contribution to victory, WAAAF members were rapidly returned to civilian life and the service was disbanded in December 1947.

WOMEN'S ROYAL AUSTRALIAN AIR FORCE

In the prosperous post-war years, the attractions of civilian employment resulted in the RAAF being unable to recruit sufficient males to fill its training courses, leading to critical undermanning in some areas. At the same time, operational demands on the RAAF were increasing with squadrons deployed on combat operations in Malaya and Korea. Recruiting females was seen as a way out of this situation.[46] The Women's Royal Australian Air Force (WRAAF) was formed in 1950, and the first recruits commenced training on 30 January 1951.

Unlike the wartime WAAAF, the WRAAF was a permanent organisation, and therefore much effort was devoted to giving it an appropriate uniform and identity. In 1950, uniforms were designed for the WRAAF by the Commonwealth Government Clothing Factory in consultation with Miss Rita Findlay, a Director of Georges' Ltd, Melbourne and, despite a certain amount of criticism, these uniforms were a big advance on wartime WAAAF issues.[47]

Service Dress – Winter

The WRAAF winter service dress uniform was a complete redesign from the uniform worn during wartime by WAAAF members, although it was still in the RAAF's dark blue colour. The tunic was worn without a belt and the skirt had a single box pleat. Buttons were black initially but, like the RAAF service dress uniform, they were changed to gilt in 1954.

Unlike the wartime WAAAF uniform, the tunics worn by WRAAF members followed the female dress convention of buttoning right-over-left.

In 1963, the winter service dress was revised with a new style of tunic. A new light blue blouse with a dark blue stripe at the peaks of the collar was worn without a tie. Also, an 'air hostess' style of hat replaced the field service cap.

The uniform was further updated in 1972.

A WRAAF section officer in winter service dress, 1953

The post 1972 WRAAF uniform. Source: RAAF Museum

---

[46] Alan Stephens, *Going Solo, The Royal Australian Air Force 1946–1971*, Australian Government Publishing

[47] Royal Australian Air Force Association, WRAAF Branch website, *History of the Women's Royal Australian Air Force 1951–1977*, 2009, at http://www.wraaf.org.au/History.html viewed 1 March 2018

The 1963 service dress uniform

Service Dress – Summer

When the WRAAF was formed in 1951, the summer service dress was the same long-sleeve blue-grey dress with black buttons and belt buckle as worn by WAAAF nursing orderlies and illustrated earlier in this chapter. It was usually worn with the khaki fur-felt hat, stockings and black shoes.

The long sleeved summer uniform was replaced in 1956 by a blue-white short-sleeved frock. Initially, the buttons were black but were later changed to gilt. In 1963, a polyester dress jacket was introduced for wear with the summer uniform frock. It was available only to officers and warrant officers at their own expense.

Summer frock introduced in 1956

WRAAFs in summer service dress in 1967

The WRAAF summer uniform included a dark blue cardigan for wear in cool weather. A new version of the cardigan was introduced in 1971 in a lighter shade of blue.

The 1971 cardigan illustrated here has fittings for shoulder boards which means this example was worn by an officer or warrant officer.

Mess Dress

Mess dress for female officers was introduced in 1965. It was a dark blue synthetic dress with white collar, matching belt and anodised buttons. Rank insignia was gold rank braid worn on shoulder boards. In 1969, this form of mess dress was extended to female SNCOs from sergeant to warrant officer.

The uniform was further updated in 1982.

Working Dress

From 1951 until well into the 1970s, the khaki dress with buttons from the collar to the hemline was worn as working dress. It was normally worn with the khaki fur-felt hat, stockings and black shoes. In cool weather, a dark blue cardigan was worn over it.

Mess dress for female officers introduced in 1965.
Source: RAAF Museum

WRAAF khaki working dress uniform from the 1950s to the 1970s

Rank Insignia

The rank structure for the WRAAF was as given in the WAAAF table earlier in this chapter with the exceptions that the highest WRAAF rank was Group Officer and Under Officers were called Warrant Officers. Rank insignia and wearing conventions were the same as the RAAF.

Headdress

When the WRAAF was formed, two forms of headdress were worn, the dark blue field service cap with winter service dress and the fur-felt hat in grey for summer dress and in khaki for working dress. Other caps were introduced later. All caps were fitted with an embroidered cap badge from 1951 to 1955. A bullion version of the embroidered badge was also produced but never issued, probably because it was only ready around the time of the death of King George VI and it would have had to be replaced with a St Edward's crown version. The embroidered WRAAF cap badge was replaced, as it became available in 1954/55, by gilt metal St Edward's crown badges, and oxidised for the khaki fur-felt hat.

WRAAF khaki fur-felt hat worn with working uniform from 1951 to 1967

WRAAF grey fur-felt other ranks' cap dated 1953, with Tudor crown embroidered badge

WRAAF grey fur-felt officer's hat dated 1954, with St. Edward's crown bullion badge

WRAAF grey fur-felt other ranks' cap dated 1954, with St. Edward's crown gilt badge

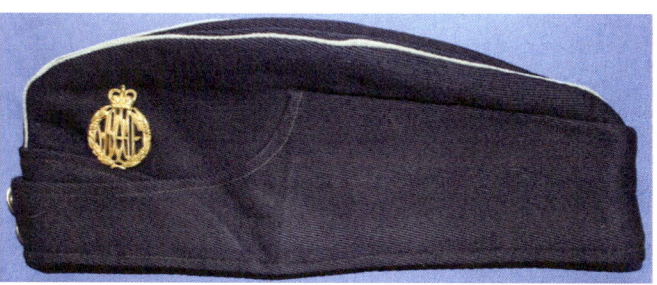

WRAAF field service cap with light blue piping, anodised 1960s crown and eagle buttons, and gilt WRAAF badge

The 'air hostess' style hat introduced with the 1963 WRAAF uniform to replace the field service cap

WRAAF cap for service and working dress introduced in 1967

WRAAF service dress cap introduced in 1969. This example with a Warrant Officer's badge

Qualification Badges

WRAAF members wore the same qualification badges as the RAAF members. See Chapters 1 to 3 for images of these badges.

Other Badges

Like RAAF members, non-commissioned members of the WRAAF wore the eagle badge on both sleeves of the tunic, just below the shoulder seam. A similar badge in khaki was worn on the sleeve of khaki summer uniforms.

When the eagle insignia on all RAAF badges changed to the Australian wedge-tailed eagle in the mid-1960s, the shoulder eagle badge worn on airmen and airwomen's winter uniforms was updated.

The eagle badge, based on the Australian wedge-tailed eagle, worn after 1966

Accoutrements

A minor but popular addition to the uniform in the mid-1950s was a black leather, sling shoulder bag. Permission to fold the strap and carry it as a handbag over the wrist was given in 1960.

The tan leather gloves which were worn with service dress were changed to black leather from the mid-1950s. Fabric gloves for summer were approved in 1966.[48]

Integration

On 1 May 1977, the WRAAF was integrated into the RAAF in keeping with the Australian Government policy of equal opportunity for females. This greatly expanded the employment opportunities for females as all RAAF musterings (trades) and categories were then open to females, except for those with a direct combat role such as aircrew and airfield defence guard.

At integration, WRAAF officers adopted RAAF officer rank titles and the WRAAF service dress cap badge was replaced by the airman's cap badge. The uniforms worn by female members after 1977 are covered in chapter 6.

---

[48] Royal Australian Air Force Association, WRAAF Branch website, *History of the Women's Royal Australian Air Force 1951–1977,* 2009, at http://www.wraaf.org.au/History.html viewed 1 March 2018

CHAPTER 5:
THE POST-WAR PERIOD, KOREA & VIETNAM, 1948–1972

With the cessation of hostilities around the world in August 1945, the RAAF's major task was to reduce in size to one that met the Government's directives for peacetime. Aircraft, equipment and facilities were sold off or destroyed while the huge task of returning personnel to Australia and discharging them took several years. Of the 173,622 members of the RAAF on the last day of the war, only 13,238 were still in the service on 31 October 1946. This number fell to just 7,897 by the end of 1948.[49]

Despite the return of peace and the massive reduction in resources, the workload of the RAAF was still substantial. Three flying squadrons and support units were sent to Japan as part of the British Commonwealth Occupation Forces (BCOF). Just as the last of these squadrons was preparing to return to Australia in 1950, North Korean forces invaded South Korea and No 77 Squadron began their three years of combat operations in support of United Nations forces. Other RAAF units continued in Korea until April 1955.

RAAF Cold War deployments did not end with the Korean War. In 1950, the RAAF squadrons commenced operations in South East Asia that would last for decades. An RAAF bomber squadron and a transport squadron deployed to Singapore in June 1950 to support Commonwealth forces fighting communist guerrillas in Malaya. By the time the bomber squadron was withdrawn, the RAAF had built a major air base at Butterworth and based three flying squadrons there. Two fighter squadrons deployed to Malta in the Mediterranean in 1952 to counter Soviet forces in Europe. In mid-1962, a squadron of Sabres was deployed to Ubon, Thailand after communist forces had come dangerously close to the Laotian/Thai border; and from 1963 to 1966, RAAF forces at Butterworth were on standby to deter Indonesia forces from destabilising the newly created nation of Malaysia.

In 1964, a squadron of Caribou transport aircraft deployed to Vietnam in what was to become the RAAF's largest wartime operation since World War II. A helicopter squadron and a bomber squadron were to follow while RAAF Hercules transport aircraft maintained the logistics bridge to Australia. The last RAAF unit only withdrew from Vietnam in 1972.

WINTER SERVICE DRESS

The year 1948 was the start of the post-war RAAF, the year in which recruiting started up again and the year that the first post-war training courses were run. It was also the year that wartime personnel policies, including the regulations on the wearing of uniforms, were changed to those more appropriate for peacetime. Wartime regulations that prohibited the wearing of ceremonial uniforms and mess dress were relaxed. Wartime austerity measures were cancelled and RAAF members could now wear civilian clothes when off-duty.

Initially, uniforms remained the same with changes being introduced slowly over many years. Officer rank braid, which had changed from gold to the black/blue lace at the start of World War II, remained black/blue after the war. The eagle and crown badge which had been mounted just above officers rank braid before the war had disappeared. Officers and warrant officers rank insignia remained on the lower sleeve while members from flight sergeant to aircraftman still wore their rank on the upper sleeve. In 1954, the RAAF, together with the WRAAF, changed their black buttons on winter service dress to gilt, although black buttons remained on battle dress and the greatcoat.

---

[49] Dr Alan Stephens, *The Australian Centenary History of Defence, Vol II, The Royal Australian Air Force*, Oxford University Press, Melbourne VIC, 2001, p 179

A greatcoat was worn during very cold conditions with winter service dress. It was the same style as that worn in the 1920s and 1930s and was made from 30-ounce woollen cloth. In 1961 the woollen material was changed to a twill but the style remained essentially the same. The 1961 greatcoat remained in use after the all-seasons uniform had been introduced.

Greatcoat from the early 1950s with black buttons

Winter service dress tunics from the 1960s. Group Captain (left) and Corporal (right)

SUMMER SERVICE DRESS

The khaki summer service dress of cotton long-sleeved shirt and trousers with a black tie was retained after the war. Officers and warrant officers sometimes wore a khaki tunic as well. The major change to this uniform was in 1960 when the cotton material was replaced by polyester, making the uniform easier to maintain.

Officer's service dress tunic with gunner's flying badge

An officer inspecting a flight of enlisted personnel in the early 1960s. Note that some of the personnel are wearing cotton service dress and some polyester, with the polyester items appearing slightly darker

CEREMONIAL DRESS

Ceremonial dress in this period was usually service dress with medals and sword for officers and warrant officers. For ranks from flight sergeant to aircraftman, ceremonial dress consisted of service or battle dress with ceremonial white belt and gaiters. If a rifle was carried, it had a white sling.

Two apprentices on sentry duty at the Apprentice Squadron, School of Radio. They are wearing service dress caps, War Service (Battle) Dress uniforms with a white webbing belt, scabbard and gaiters. The weapon is the ubiquitous SMLE .303 rifle with bayonet

TROPICAL DRESS

With many units based in tropical areas in Malaysia, Singapore and northern Australia, tropical dress was a practical and popular uniform. Like the summer service dress uniform, tropical dress was made from cotton material in the 1950s, changing to polyester in 1960. This uniform remained substantially unchanged until it was withdrawn from use in 2013.

Tropical dress or 'drabs' in the 1960s

MESS DRESS

Mess dress was re-introduced in 1949.[50] After the war, much of the formality of mess life had disappeared and a simplified version of officers mess dress reflected this. As mess undress was no longer in use, only two versions of mess dress were worn—one for winter and a lighter one for summer. Winter mess dress consisted of an Eton-style dark blue short jacket with matching trousers, a white shirt, and black bow-tie. A dark blue cummerbund around the waist replaced the waistcoat. The principal change from the pre-war design was that the silk facing of the jacket was changed from light blue to RAAF dark blue. Miniature medals and qualification badges were worn on the left lapel. Rank braid remained in gold but was worn on shoulder boards instead of the sleeve.

The wearing of mess dress was extended to warrant officers, flight sergeants and sergeants in 1968. This uniform consisted of dark blue trousers, a long-sleeved white shirt (later to be changed to a Marcella fronted shirt), a black bow-tie and a white waist-length jacket similar to the officers' summer jacket, albeit without provision for shoulder boards. A dark blue cummerbund was worn around the waist of the trousers. Gold bullion rank insignia was worn on the right sleeve initially, but later on both sleeves.

---

[50] Air Board Order N.1 dated 21 January 1949.

A uniform similar to the SNCOs mess dress was also worn by stewards in officers' and sergeants' messes.

Winter mess dress from the mid-1950s, shown with Flight Lieutenant rank and signaller brevet

Summer mess dress from the early 1950s with Air Commodore rank and pilot brevet

WORKING DRESS

War Service Dress or Battle Dress

War service dress, or battle dress as it was generally known, continued to be worn after the war as a working dress in winter. Medal ribbons and qualification badges were worn on the left breast. In 1961, the serge material of battle dress was replaced by the finer twill material.

Pilot 3 G.I. Stephens wearing battle dress uniform with aircrew 3 rank insignia and pilot brevet in the late 1940s (Source: AWM Canberra)

Jungle Greens

The working dress uniform known as 'jungle greens' had its origins in New Guinea during World War II, where the Australian Army and Air Force troops changed their khaki uniforms to dark green to blend into the green vegetation of the tropical jungle. By October 1942, jungle green garments were being shipped to Army units in New Guinea and became the standard working/combat dress for Army personnel in tropical areas. In October 1965, the RAAF created the new mustering of airfield defence guard (ADG) to provide ground defence for its base at Butterworth, Malaysia and other forward airfields. Jungle greens were made the working dress of the new mustering because of the combat-related nature of their training and duties. Their clothing and ancillary equipment was the same as that of the Australian Army, except for the insignia. All appropriate rank and other insignia were made in jungle green material to match the uniform. Jungle greens also became associated with ADGs carrying out their combat role of defending airfields in Vietnam.

The basic jungle green working uniform consisted of cotton trousers and long sleeve shirt. Additional cold weather items were a 'Howard Green' pullover and a 'combat smock'. This was a long jacket with a waterproof, polyester outer layer and an inner, detachable liner containing Dacron material. Warrant Officers and below wore their rank insignia on the sleeves of the shirt and pullover while officers wore jungle green rank slides attached to the epaulettes of the shirt. Black general purpose boots were the authorised footwear. Headdress in the field was usually the cotton utility hat of the same colour as the clothes, but the fur-felt hat could also be worn.

RAAF Airfield Defence Guards at Phang Rang, Vietnam in 1971

An example of a Jungle Greens rank slide

RANK INSIGNIA

Officer rank insignia in use after World War II remained the same as that used during the war. Rank braid used on service dress tunics remained the black/blue lace and did not return to the gold braid that had been used before the war.

The only rank slides in use in this period were khaki officer and warrant officer slides for wear on summer service dress and tropical dress shirts. The cloth varied from fine cotton to cotton drill and came in varying shades of khaki. Dark blue rank slides were not needed as service dress tunics and battle dress had the rank braid sewn directly on. The winter light blue shirt had no facility for rank insignia as the shirt was never worn as an outer garment.

Shortly after the end of the war, the metallic coat-of-arms badge worn by warrant officers on the sleeve of winter service dress tunics and battle dress blouses changed from oxidised (black) to anodised (bright).

In 1966, nearly 30 years after the first attempt, approval was given to replace the Royal Air Force (RAF) eagle on RAAF uniforms with the Australian wedge-tailed eagle. See later section on Headdress for the full story. The change of eagles included those on the eagle-and-crown badges worn by officers and warrant officers.

Summer dress rank slides. Top: Pilot Officer, Flying Officer, Flight Lieutenant, Squadron Leader, Wing Commander. Bottom: Group Captain, Air Commodore, Air Commodore (overseas), Air Vice Marshal

Warrant Officer rank insignia followed the same post-war pattern of changes from King's crown to Queen's crown with a further change from the Royal Coat of Arms to the Australian Coat of Arms in the 1960s. The anodised sleeve badge manufactured by Stokes and shown on the following page was made without lugs and was designed for sewing onto the sleeve. Examples are sometimes found with added pins which have been adapted by the owners.

Warrant Officer rank slides showing the oxidised eagle with King's crown to 1954, the oxidised eagle with Queen's crown from 1954–1966, and an anodised Australian wedge-tailed eagle worn from 1966

WOFF Post-war, gilt, KC to 1954

WOFF Post-war, anodised, QC (Stokes, Melbourne)

WOFF Post-war, anodised, QC, from 1954

WOFF Dark blue, QC, from 1954

WOFF Khaki drabs, QC, from 1954

WOFF Dark blue, 1960s Commonwealth Coat of Arms

Another change to rank insignia after World War II was the use of smaller chevrons in the rank badges for flight sergeants, sergeants and corporals. For these ranks, the complete badges were manufactured rather than the previous style where chevrons were cut from a roll of stock, as required, and a separate embroidered or gilt metal crown was mounted above the chevrons for flight sergeants.

Flight Sergeant

Sergeant

Corporal & LAC

The leading aircraftman (LAC) badge was unchanged from the wartime period.

The rank badges were also made in khaki for use on summer and tropical uniforms.

Flight Sergeant   *Sergeant*   *Corporal & LAC*

Cadet Rank Insignia

Since the inception of the RAAF, cadet aircrew were distinguished by a white band on their service dress caps and, white gorgets at the lapel step openings of the service dress jacket. When wearing battle dress or a drab uniform without jacket, cadets wore an all-white slide on their shoulder straps.

When the RAAF College was established at Point Cook on 1 August 1947, the length of the course was four years, considerably longer than the one year that most non-College trainee aircrew spent on their training courses. There was a need for cadets at the College to wear some insignia to indicate their relative status. Warrant officer, flight sergeant and sergeant insignia were worn on the sleeve in addition to the white gorgets to distinguish cadets by study year and nominal (unpaid) positions in the cadet hierarchy.

An RAAF College graduating class in the 1950s, with cadets wearing SNCO rank insignia together with white cap bands and gorgets

  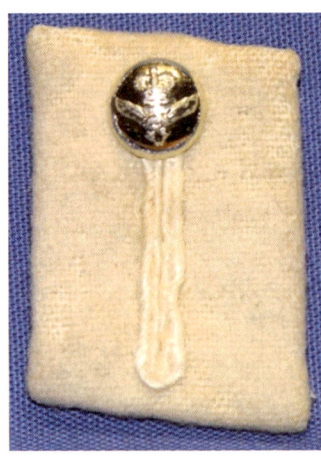

Officer Cadet gorgets with black button (WW2), gilt button (1950s) and anodised button (1960s)

By 1960, RAAF College cadets wore white shoulder slides on the epaulettes with bands to indicate seniority. When the Diploma Cadet Squadron began courses to train engineer officers in 1962, the cadets wore the same cadet shoulder slides.

Cadet   *Air Cadet*   *Senior Air Cadet*

Air Cadet Under Officer   *Squadron Cadet Under Officer*

Apprentice Insignia

In 1948, the first group of apprentice trainees (aged 15–17 years) commenced their training at RAAF Forest Hill, NSW (near Wagga Wagga). A short while later, a further group of trainees commenced training at the Radio Apprentice School at RAAF Frognall in the Melbourne suburb of Canterbury. The Apprentice Scheme came to an end in 1993, by which time more than 6,150 personnel had been trained in various technical trades.

The apprentices were issued with their own identifying insignia, which comprised a light blue mohair cap band (later with a light blue disc behind the cap badge, see page 112) on their service dress caps and a blue triangular material badge worn on the upper sleeve of jackets in lieu of the eagle. In summer dress, a blue triangle was sewn onto the rank slides worn on the epaulettes of the drab summer shirt.

The apprentice shoulder badge (left) worn on winter service dress and the apprentice rank slide (right) worn on khaki summer uniforms in the 1950s

Later, the blue triangle badge was replaced by an apprentice shoulder flash which incorporated the eagle. After 1966, the winter service dress blue version was updated with the wedge-tailed eagle but there is no evidence of the khaki badge being updated.

Pre-1966 Apprentice shoulder badge with RAF type eagle

1966 Apprentice shoulder badge with Australian eagle

Apprentice badge for summer & tropical dress

1966 badge with red line for a leading apprentice

RAAF Trainees

RAAF trainees were non-commissioned trainees who were older than 18 years, later reduced to 17. Photographic evidence suggests that in the 1950s, trainees wore a white tape along the length of the epaulettes of their khaki shirts and battle dress blouses.

Members wearing this insignia were undergoing their initial recruit training at the time and held the rank of aircraftman recruit (ACR).

A young airman in the 1950s wearing khaki service dress. The white tape on the epaulettes indicated he was a trainee

Non-Commissioned Aircrew Rank Badges

With the first post-war aircrew course starting in 1948, the RAAF introduced a rank system where non-commissioned aircrew held specific aircrew ranks. The system was similar to one used by the RAF. Aircrew recruits would be known as 'aircrew trainee' and, once qualified, their rank would include their specialisation, followed by a number to indicate their experience and status. For example, a newly graduated navigator would hold the rank of Navigator 4 and he would be promoted over many years to Navigator 1. Select aircrew could be promoted to 'aircrew master'. At any time after graduation, an aircrew member could be commissioned and would then hold officer rank.[51]

For legal status and privileges, the aircrew ranks were given equivalent status to existing RAAF ranks as shown.

Aircrew rank	Equivalent rank
Aircrew Trainee	Aircraftman
Aircrew 4	Leading Aircraftman
Aircrew 3	Corporal
Aircrew 2	Sergeant
Aircrew 1	Flight Sergeant
Aircrew Master	Warrant Officer

The rank badges for the new system were copied from the RAF system and featured stars indicating the seniority and status. The badge for aircrew master featured the coat-of-arms at the top of the wreath. The badges were produced in opposing pairs, on blue melton cloth and in khaki drab versions for winter and summer uniforms respectively.

Aircrew Master *Aircrew 1* *Aircrew 2*

Aircrew 3 *Aircrew 4* *Aircrew Trainee*

---

[51] Alan Stephens, *Going Solo, The Royal Australian Air Force 1946–1971*, Australian Government Publishing Service, Canberra ACT, 1995, p93

The system, colloquially known as the 'star and garter' system, was not popular. In 1950, it was replaced with the conventional rank system where newly graduated aircrew had the rank of sergeant and were promoted to flight sergeant and warrant officer over time.[52]

The aircrew trainee badge was the only badge of this rank system which remained in use after 1950. It was worn by trainees until the early 1960s.

Aircrew Trainee

Senior NCO Mess Dress Insignia

Senior NCOs wore rank insignia in bullion and gold braid on a white background on their tunics. Mess stewards wore a jacket similar to the SNCO jacket; corporals and leading aircraftmen catering personnel wore rank insignia of the same gold braid design on their white jackets.

Warrant Officer   *Flight Sergeant's Crown*   *Sergeant*   *Corporal (Catering)*

HEADDRESS

Like other forms of dress, hats and caps initially stayed the same as those worn during wartime but the following changes occurred over the next three decades. Most of the changes were to the badges rather than to the cap or hat.

Soon after the war, the 'economy' officer cap badge was replaced with the bullion badge that had been used before the war. The airman's cap badge reverted to the gilt finish from the wartime black oxidised finish.

The khaki cover worn over the top of the service dress cap with khaki uniforms ceased to be used from the early 1960s.

---

[52] Alan Stephens, *Going Solo, The Royal Australian Air Force 1946–1971,* Australian Government Publishing Service, Canberra ACT, 1995, p 95

The oxidised badge containing the Tudor crown marks this airman's cap as coming from the late 1940s period

A 1950s officer cap with bullion Tudor crown badge

The service dress cap of an Air Vice-Marshal from the early 1950s

1950s Airman's caps. Above: Summer wear with oxidised Tudor crown and khaki cover. Below: Winter wear with the later St. Edward's crown

Change from Tudor Crown to St Edward's Crown

The crown in RAAF badges is a symbol of the reigning monarch. Since the creation of the RAAF in 1921, the Tudor crown was used in heraldry as the symbol of the English kings George V, Edward VIII and George VI. When Elizabeth II acceded to the throne in 1952, she chose the St Edward's crown as her heraldic symbol. By 1954, all new RAAF badges including cap badges, eagle and crown badges, buttons bearing a crown and flying badges incorporated the St Edward's crown. Badges with the Tudor crown continued to be worn after 1954 until they were replaced.

Airman's gilt cap badge with the Tudor crown

Airman's gilt cap badge with St Edward's crown, 1954-66

Warrant Officer's cap badge, St Edward's crown, 1954-66

Change from RAF Eagle to Wedge-Tailed Eagle

In 1937, the RAAF Air Liaison Officer in London was instructed by the Air Board to approach the College of Heralds to design a badge for the RAAF along the lines of the Royal Air Force's badge but incorporating the Australian wedge-tailed eagle. The Chester Herald, being the Inspector of RAF badges, was only too happy to comply with this request, pointing out that the Royal Air Force 'eagle' was often mistaken for an albatross. Progress was made and, by October 1938, the Air Board was shown the design for an RAAF badge with an Australian eagle. In due course, Royal Assent would be sought for the badge.

Then in January 1943, the Chester Herald reported an ornithological error in his original drawing. The eagle's two central tail feathers were not long enough. The original heraldic painting, on vellum, was returned to London for correction by the Chester Herald. With the war in Europe pressing, the matter of the badge was put on the back burner.

Finally, in 1966, the matter was raised again and approval was granted for the change of eagle on RAAF badges. The first cap badge to be produced in the new style was that of a warrant officer, with those for air officers and officers to follow. Cap badges with the pre-1966 eagle continued to be issued into the 1970s until stocks ran out.

The eagle-and-crown badges used on rank slides and shoulder boards demonstrate the difference between the RAF eagle on the left and the Australian wedge-tailed eagle on the right

The post-1966 service cap badges for an air officer (left) and an officer of Group Captain and below (right) with the Australian wedge-tailed eagle

A late 1950s/early 1960s service dress cap fitted with a St Edward's crown officer's cap badge and cadet's white fabric hat band

Apprentice's service dress cap worn with the light blue cap band and a coloured disc behind the cap badge. Dates from 1966

Field Service Cap

The field service cap continued to be worn until 1952, although medical personnel on aeromedical evacuation duties continued to wear it until at least 1956. The black oxidised cap badges worn in wartime reverted to the gilt finish about 1948.

Field service cap with officer's and warrant officer's gilt badge from the late 1940s

Fur-Felt Hat

The fur-felt hat continued as the form of headdress with tropical dress, virtually unchanged from its wartime form. The cap badge changed later to the bright anodised finish.

Fur-felt hat with oxidised airman's cap badge and puggaree

QUALIFICATION BADGES: AIRCREW

In 1948, the flying badges in use were pilot, navigator, signaller, air gunner and flight engineer. Due to wartime austerity, the only form these badges took was embroidered white and blue silk on a black background. Only the pilot brevet was surmounted with a crown; the other badges consisted of a blue wreath with a letter monogram in the centre and a single outstretched wing on the right side. Over the next three decades, the following changes to flying badges took place.

In January 1949, the badges for navigator, signaller, air gunner and flight engineer were redesigned to add a crown to the top of the badge. In accordance with Royal protocols at the time, this crown was a Tudor crown. The air gunner badge was also renamed 'gunner' and the monogram was changed from 'AG' to 'G'.[53]

Pilot brevet World War II to 1954

Navigator from 1949 to 1954

Signaller from 1949 to 1954

Flight Engineer from 1949 to 1954

Gunner from 1949 to 1954

---

[53] Air Board Order N.2 dated 21 January 1949

Sterling Silver Flying Badges

Sterling silver flying badges were first introduced into the RAAF in 1949 for wear on khaki drab (summer) shirts and jackets.[54] Like the new embroidered flying badges, all sterling silver badges were surmounted by a Tudor crown. They were made with a brooch fitting to enable easy removal from the shirt when washing was required. The initial 1949 production run was contracted to Stokes & Sons with a supplementary production run by K G Luke in 1952. All five Tudor crown badges are found embossed STG SIL with a makers' mark, 'Stokes & Sons 1949' or 'KG Luke Melbourne 1952', on the reverse.

The original sterling silver pilot brevet issued in 1949 to Flight Lieutenant N. McNamara, later Air Chief Marshal Sir Neville McNamara KBE AO AFC, Chief of Air Staff 1979–1982

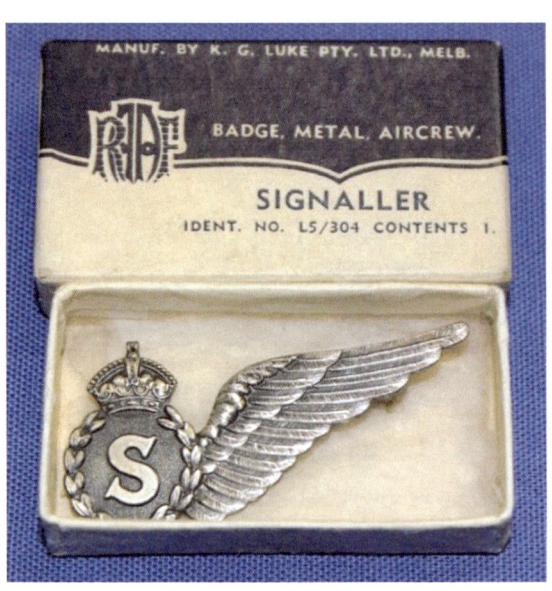

A 1952 manufacture Sterling Silver Signaller in K.G. Luke box of issue

Tudor crown flying badges in sterling silver for Navigator, Flight Engineer and Gunner worn from 1949 to 1954. The reverse shows an example of the maker's mark for Stokes & Sons 1949

---

[54] Air Board Order N.2 dated 21 January 1949

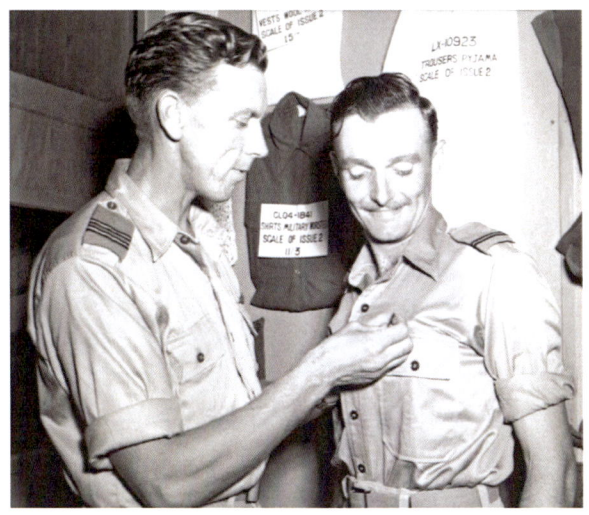

Two RAAF pilots who were shot down and held as prisoners-of-war in North Korea were repatriated to Iwakuni, Japan in September 1953 and issued replacement uniforms. Flight Lieutenant Gordon Harvey fixing a new sterling silver pilot brevet to the shirt of Flying Officer Ron Guthrie.
Source: AWM Canberra

Mess Dress & Full Size Bullion Flying Badges

At the same time as mess dress was re-introduced in 1949, miniature flying badges became available for wear on this form of dress.

Mess Dress miniature flying badges from 1949 with the Tudor crown

Although re-introduction of full-size bullion flying badges was approved in the orders of 1949, it appears that the first of these did not come into use until 1954. They were produced for all the aircrew categories with a St Edward's crown for wear on the winter service dress tunic.

Bullion pilot brevet with St Edward's crown from the late 1950s

Change to St Edward's Crown

In 1954, as with other RAAF badges, the design of the flying badges was changed to replace the Tudor crown with the St Edward's crown.

Embroidered & sterling silver flying badges updated with the St Edward's crown from 1954

In the post-1954 era, reorganisation and developing technology led to a number of revisions and reclassifications of musterings in the RAAF. See Appendix 10 for different makers of each wing.

Parachute Jump Instructor Badge 1951-1972

In the post-war decades, the RAAF was responsible for all parachute training. In 1951, a parachute jump instructor badge was created to recognise the skills of these members who regularly flew in transport aircraft while supervising parachuting operations. The initial badges were manufactured in silver plated or silver washed bronze or brass, without the crown. The crown was added in 1954 when all flying badges were redesigned to incorporate the St Edward's crown. The badge was no longer issued after responsibility for parachute training was transferred to the Army in 1972.

Uncrowned cloth PJI badge 1951-1954

Uncrowned PJI badge 1951-1954

Crowned cloth PJI badge 1954-1972

Crowned sterling silver PJI badge 1954-1972

Loadmaster Brevet 1963

When the C-130 Hercules transport aircraft was introduced in 1958, it required a specialist on the crew to handle the cargo/personnel loading and unloading. Initially, this crew member was identified by a scarlet brassard but this was replaced by the loadmaster brevet in 1963. This brevet was later worn by loadmasters on other transport aircraft such as the C-47 Dakota and the Caribou.

Loadmaster embroidered and sterling silver badges from 1963

Air Electronics Officer Brevet 1965-1979

The introduction of the SP-2H Neptune aircraft in 1962 showed the need for a more highly trained crewmember to operate the sophisticated electronics equipment on-board and led to the introduction of the air electronics officer (AEO) category on maritime patrol aircraft. The air electronics brevet (AE) was first awarded in December 1965 and continued until AEO training ceased in 1979. Once AEO training commenced in 1965, no more signallers were trained and the signaller brevet was no longer awarded.

Air Electronics Officer embroidered and sterling silver badges from 1965

Crewman Brevet 1966

When the Air Force began operating Iroquois helicopters in the 1960s, a non-pilot crewmember was found necessary to assist with the loading/unloading and to keep a lookout underneath and behind the aircraft. The first crewman (C) brevets were awarded in 1966.

Crewman embroidered and sterling silver badges from 1966

Re-introduction of the Gunner (G) Brevet 1966

The gunner brevet was introduced in 1940 for the air gunners on bomber and maritime aircraft. In the post-war period, gunners continued to fly on Lincoln bomber aircraft but, when the last of these were replaced by Canberra bombers in 1961, the gunner role ceased and the brevet was no longer awarded.

In 1966, Iroquois helicopters flying in the Vietnam War had two door guns—one on each side of the aircraft. The crewman was trained to operate one gun but for wartime flying, an assistant crewman was required to operate the other gun. Most personnel employed in this role were airfield defence guards because they were experienced in using machine guns. The Gunner brevet was reintroduced to identify them as aircrew. After No 9 Squadron withdrew from Vietnam in 1971, the need for a gunner on Iroquois aircraft no longer existed although the brevet was still being awarded in the 1990s.

QUALIFICATION BADGES: GROUND STAFF

Medical Branch Badge

Since 1937, the RAAF had used the caduceus of Mercury—a staff entwined with two serpents—as the symbol of the medical branch, similar to the Royal Air Force. On closer scrutiny of mythology in the 1960s, an error was discovered. The Roman god Mercury, who was the messenger of the gods and the patron of merchants, travellers and thieves, had been confused with the Greek god Aesculapius who was the god of healing and medicine. The symbol of Aesculapius was a rod entwined with only one serpent.[55] Many people attribute the error to the US Army Medical Corps which adopted the caduceus as its symbol in 1902.[56]

In the 1960s, the RAAF decided to correct the error and adopt the symbol of the Greek god of medicine as the basis for the badge of the medical branch. This featured only one serpent under the wings and crown. The RAAF also adopted the Latin spelling of Aesculapius. The badges are illustrated under Medical Branch in Chapter 4.

OTHER BADGES AND INSIGNIA

Shoulder Devices

After the war, the wearing of shoulder devices by the holders of honorary positions to the reigning monarch resumed. The devices were worn on the shoulders of service dress, ceremonial dress and mess dress.

With the accession of Queen Elizabeth II to the throne in 1952, the royal cypher changed to EIIR and the shoulder device was changed accordingly.

Citizen Air Force Badge

Prior to World War II, the Citizen Air Force (CAF) consisted of members serving on a part-time basis, such as weekends and was intended as a backup to the Permanent Air Force. During the war, CAF members served fulltime, so the CAF as a reserve force effectively ceased to exist. At the end of the war, former CAF members were discharged and the force was disbanded.

On 1 April 1948, the CAF was reformed and four reserve squadrons were created. CAF members again served on a part-time basis and wore the same uniforms as Permanent Air Force members but were identified as reservists by the CAF badge worn on each shoulder. This badge continued to be worn until the 1980s when the CAF was reorganised into the RAAF Active Reserve. There were a number of manufacturing variations over this period, as illustrated on the following page.

---

[55] Walter J Friedlander, The Golden Wand of Medicine: A History of the Caduceus symbol in medicine. Greenwood Press, Santa Barbara CA, 1992

[56] Wikipedia, subject Rod of Asclepius, viewed 1 February 2018 at en.wikipedia.org/wiki/Rod of Asclepius

CAF insignia with oilskin backing   *Heavily padded*

Padded on felt backing   *Flat with felt backing*   *Flat with cotton backing*

Shoulder Eagles/Badges for Personnel Serving within Australia

Non-commissioned members of the RAAF continued to wear the eagle badges illustrated in Chapter 2 (RAF type, until 1966) and Chapter 4 (wedge tailed eagle, from 1966) on the shoulders of both sleeves of the tunic. The khaki version also continued to be worn on khaki summer uniforms.

'RAAF' Badges

It was important to distinguish members of the RAAF wearing khaki uniforms from members of the Army. For officers and warrant officers, 'RAAF' was embroidered on the rank slides worn on the epaulettes. Members from aircraftman to flight sergeant wore a small khaki slide with 'RAAF' embroidered in red cotton which was changed to blue in the mid-1960s. When serving overseas, the slide was changed for one with 'AUSTRALIA' on it.

Country of Origin Badges

RAAF personnel serving outside of Australia wore a badge on each shoulder to identify their country of origin. Officers and Warrant Officers wore the 'AUSTRALIA' shoulder badge whereas airmen from aircraftman to flight sergeant wore a combined 'AUSTRALIA' over eagle badge.[57] The badges were blue on black for winter dress and brown on khaki for summer and tropical dress.

---

[57] Air Force Order 5/D/9 (W.O. 1000) dated 27 July 1953

Country badges worn by Officers, Warrant Officers and NCO aircrew

Country badges worn by non-aircrew airmen

Badges of the RAAF Force in Japan and Korea

With the end of World War II in August 1945, the British Commonwealth Occupation Force (BCOF) was formed to enforce, along with US forces, the terms of the unconditional surrender of the Japanese. The BCOF was based in southern Japan and the RAAF component consisted of three fighter squadrons. As the role of the force was diminishing, the Korean War broke out in June 1950 and several BCOF units deployed to the Korean Peninsula, including No 77 Squadron RAAF. From November 1951, the British Commonwealth Forces Korea (BCFK) superseded the BCOF and commanded all Commonwealth forces in Korea and Japan.

Three types of distinctive badges were issued to personnel of all services, for wear on their uniforms during the various phases of the operations. These were worn on the right upper sleeve of greatcoats, tunics, battle dress blouses and shirts, one inch below the country-of-origin badges. As with other RAAF badges, the crown on these badges changed from the Tudor crown to the St Edward's crown in 1954.

Japan 1946–1951 *Korea 1951–1954* *Korea 1954–1955*

Buttons

In the 1960s, the gilt buttons with eagle-and-crown and 'R.A.A.F.' were superseded by anodised buttons with the Australian wedge-tailed eagle. These buttons continued to be worn on service dress and mess dress into the 21st century.

Officer's Sword Belt

Since the earliest years of the RAAF, officers and warrant officers had suspended their sword scabbards from a gold brocade belt fastened by a gilt buckle. In 1954, the crown on the belt buckle changed from the Tudor crown to the St Edward's crown. With the introduction of the all-seasons uniform in 1972, the brocade belt was no longer worn. Instead, hangers to carry the scabbard were attached to the trouser belt.

Sword belt buckle with Tudor crown, worn until 1954

Gold brocade sword belt and buckle with St Edward's crown, worn after 1954

CHAPTER 6: THE 'ALL SEASONS' UNIFORM 1972–2000

In the mid-1960s, the Air Board considered a number of initiatives to mark the fiftieth anniversary of the RAAF, which was to occur in March 1971. Among these was a proposal to make significant changes to the colour and style of the Air Force uniform, which was widely supported. While minor changes to material and accoutrements had been made during preceding decades, the uniform style and colour had remained basically the same since the 1930s. Senior RAAF commanders were aware of widespread complaints regarding the current uniforms. Airmen were unhappy with the difference in quality between their winter service dress uniform compared with the much higher quality of the same garments issued to officers. Across the board, many personnel felt the dark colour of uniform showed dirt and fluff too readily. This was the bane of a trainee's life at initial training units where trainees spent considerable time brushing their dark blue uniforms every morning before parade. Others felt the summer and winter versions of the uniform, which had to be changed on a set date, often did not suit local weather conditions.

As approved by the Air Board, a project team began developing a new uniform for trial. The new colour was a lighter blue than the dark blue that had been a feature of the RAAF uniform since it was finalised in 1922. It was felt that this colour, described as 'blue-grey', was more modern and in keeping with current fashion. All uniform items that had been in dark blue were now to be made in blue-grey — tunics, trousers, hats and ties. A service dress jacket and skirt in the new colour was also to be developed for females, as well as a blue-grey 'pillbox' cap.

The Governor General, Sir Ninian Stephen inspecting members of No 2 Stores Depot on parade wearing the all-seasons uniform at Regents Park NSW in 1987

Wearer trials were conducted on Air Force bases throughout Australia and also at Butterworth, in Malaysia where the RAAF had a major base. Across the board, the results were favourable. One outfit which did not pass muster was a blue-grey version of the khaki tropical dress. This consisted of blue-grey shorts, a light blue shirt and dark blue long socks. Many felt this uniform looked 'unmilitary' and others said it made the wearer look like a postman or schoolboy! However, given the overall approval rating for the service dress version of the uniform, the Air Board approved the introduction of the new blue-grey uniform in 1970.

The blue-grey material selected was a blend of 75 per cent wool and 25 per cent polyester. It was chosen not just for its look, but also for its durability. Production commenced in the same year with delivery planned for 1971, the anniversary year. However, manufacturing delays resulted in the new items not being available until 1972. Once sufficient stocks were in place, distribution went ahead in a smooth and well-planned operation. A team of tailoring and supply staff deployed to each RAAF base. At every location, the team operated from a hangar or warehouse to kit hundreds of personnel per day with their new uniforms.

The 'all-seasons' concept of this uniform meant that, for every-day wear, the member could choose which uniform items to wear. When uniformity was required, such as on a parade, the local commander could determine the uniform that all members would wear, basing the decision on the occasion and the likely weather. For the first time since 1921, the RAAF did not lay down a date when members would change from summer to winter uniform and vice versa.

The khaki or drab colour was retained for tropical dress which members were able to wear in tropical areas or in the summer months, where appropriate.

When the plan for introducing a new uniform was made, the replacement of the male uniform was given the highest priority, while the task of redesigning the uniforms of RAAF Nursing Service (RAAFNS) and Women's Royal Australian Air Force (WRAAF) was planned for the mid-1970s. The female uniforms had been updated with new materials and styles in the 1960s (Chapter 4) while the male uniforms had undergone only minor changes since the end of World War II. The new designs for the WRAAF and RAAFNS uniforms were announced in late 1974.[58] and were scheduled for introduction in early 1975. It had proved impractical to provide the WRAAF with an 'all-seasons' uniform, but the new uniform offered scope for 'mixing and matching' items to suit the local conditions. In the case of the RAAFNS, the new uniform replaced the former summer frock and hat and the winter suit and hat.

Another event in 1977 had a major impact on females in the Service and their uniforms. In the late 1960s and early 1970s, Australian social attitudes to female employment were changing, particularly in the areas of pay and equal opportunity. Equality in the Australian Defence Force was introduced in 1977 when the RAAFNS and WRAAF were integrated into the RAAF, giving females the same pay and conditions of service as their male counterparts. Although some musterings were closed to females, the majority were now open to them, including all the aircraft technical trades.[59] Females were now not required to resign on marriage and could continue to serve when pregnant, with some restrictions based on health and safety. Uniforms appropriate for females employed in a greater range of trades were now required.

Service Dress for Males

The all-seasons uniform policy initially allowed for only two forms of service dress—one with the tunic and the other, without the tunic but wearing long sleeved shirt and tie. Later, other versions such as one with a short-sleeved shirt and no tie were authorised.

The cut of the service dress tunic was waisted rather than bloused and did not incorporate a belt. Officers wore their rank on shoulder boards of the tunic while warrant officers, airmen and airwomen continued to

---

[58] RAAF News, December 1974, p.7

[59] Dr Alan Stephens, The Australian Centenary History of Defence, Vol II, The Royal Australian Air Force, Oxford University Press, Melbourne VIC, 2001, p 207

wear their rank on the sleeve.

The gilt eagle and crown badge which had been part of the RAAF uniform since 1921 was retained on the all-seasons uniform but worn in different places. Officers wore the badge above the rank braid on their shoulder boards while warrant officers wore a shoulder board with only the eagle and crown badge. Flight sergeants and below wore the badge on both lapels of the tunic.

RAAF News Vol 13 No 7 August 1971

So that the all-seasons uniform could be worn without the tunic, several changes were made to other items. First, the light blue polyester shirt had two buttoned pockets as well as shoulder straps which secured light blue rank slides or eagle and crown badges. Second, the trousers were fitted with a belt which dispensed with the braces that had been worn with the previous dark blue version of service dress and battle dress.

The tie and socks, which previously were black, were now blue-grey. A new belt, similar in style to the khaki belt worn with summer drab uniform, but in blue-grey colour, was introduced for wear with the all-season trousers/slacks.

In 1984, shoulder boards for officers and warrant officers disappeared and officers' rank insignia returned to the sleeves of their service dress tunics. Warrant officers continued to wear their coat-of-arms badge on the lower sleeve.

Group Captain with rank on shoulder boards   *Warrant Officer with sleeve insignia*   *Flight Lieutenant with rank on lower sleeve*

Service Dress Shirts

The long-sleeved service dress shirts introduced in 1972 were of a light blue cotton/polyester material with pocket flaps and epaulettes. Officer rank was worn on shoulder rank slides of the same colour as the shirt, with the appropriate rank lace and the word 'AUSTRALIA' at the base. For warrant officers and airmen, rank was indicated by embroidered insignia on a light blue background on the sleeve. Warrant officers continued to wear plain rank slides bearing the eagle and crown insignia, while other airman ranks wore a blue 'AUSTRALIA' slide on their shirt epaulettes. By the late 1970s, a short-sleeved version of the shirt became available.

In 1986, the Queen's Colour Parade at RAAF Richmond, saw the introduction of the new blue-white shirts. Rank insignia, for all ranks, was worn on blue-grey AUSTRALIA rank slides on the epaulettes of the shirt.

Service dress for females

The change to an all-seasons blue-grey WRAAF uniform came in early 1975. The new uniform comprised tunic, skirt and slacks in the same material as that of their male counterparts. Headwear was a 2-tone hostess style hat, which was replaced in the 1980s by a blue-grey version. Like the airmen's uniform, the airwomen's tunic had the rank on the sleeves and an eagle-and-crown badge on each lapel. Female officers wore their rank on shoulder boards.

Female all-seasons service dress post-1975

Summer Dress

In 1975, the new summer dress was made in a polyester viscose, without a belt and was a departure from the military pocket style that had been introduced in 1951. The new design included a princess line shirtmaker collar neck, rounded collar finish, concealed button front opening to below waist-level and concealed slanted pocket.

In 1987, shortly after the new blue-white service dress shirt for males had been introduced, a new female summer dress was designed made from the same material as the shirt. The dress had two upper pockets fastened by gilt buttons, with larger gilt buttons securing the front opening to below waist and a light blue waist belt with gilt buckle. The rank insignia moved from the sleeve to rank slides attached to the epaulettes. In due course, a blue-grey cardigan and a V-neck jumper were introduced as components of the female uniform. Additionally, maternity uniforms were developed.

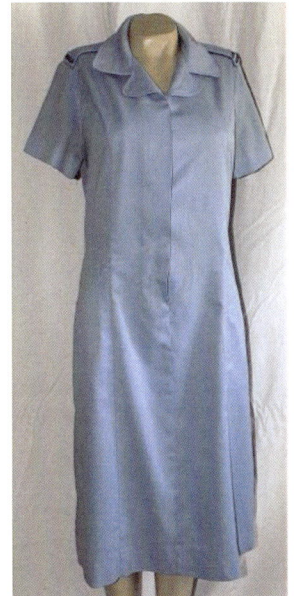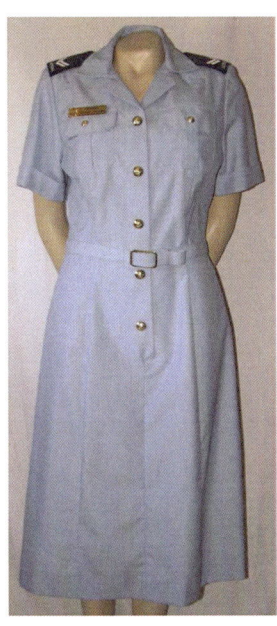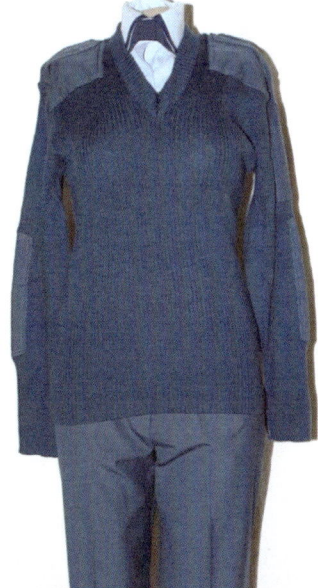

Summer dress 1974 to 1987 *Summer dress 1987 to 2000* *Service dress cardigan* *Service dress V-neck jumper*

CEREMONIAL DRESS

Ceremonial dress for the majority of RAAF members consisted of service dress (with or without the tunic) worn with medals. Accoutrements such as swords or rifles were carried as required for the occasion. Air rank officers in ceremonial dress commonly wore bullion (gold) shoulder boards, a gold sash and aiguillettes.

Ceremonial Dress

TROPICAL DRESS

Although a blue-grey version of tropical dress was trialled, it was found unsuitable and the previous khaki tropical dress was retained (Chapter 5). This was now the only khaki uniform worn by the RAAF.

MESS DRESS

Male Mess Dress

When the all-seasons uniform was introduced in 1972, no changes were made to mess dress. In winter, officers wore the dark blue Eton-style jacket with matching trousers. In summer, the dark blue jacket was replaced by a white one. Warrant officers and senior non-commissioned officers (SNCOs) wore a white Eton-style jacket with dark blue trousers all year and generally attached rank insignia on the sleeves with press studs and later with 'Velcro' patches. (Chapter 5). Bullion insignia were either sewn on or similarly attached with press studs or 'Velcro'.

After 30 June 1993, mess dress rank insignia for non-commissioned officers moved to dark blue shoulder boards, similar to those worn by officers.

Summer mess dress dated 1962 with Warrant Officer rank and Loadmaster brevet. Note that this example is fitted with the later (post-June 1993) shoulder boards

Warrant Officer of the Air Force   *Warrant Officer*   *Flight Sergeant*   *Sergeant*

Female Mess Dress

Female officers and SNCOs wore service dress or equivalent civilian attire on formal occasions such as balls and dining-in nights. Later, The RAAF introduced a female officer mess dress which consisted of a dark blue jacket and skirt worn with a white shirt and black bow-tie. Gold rank braid was worn on the sleeves.

A female Wing Commander in mess dress in the late 1970s talking with Mr D. J. (Jim) Killen, the Minister for Defence, (later Sir James Killen AC KCMG), who had served in the RAAF as an air gunner in World War II

1986 Mess dress for female officers, shown without the black bow-tie

In 1986, a new style of mess dress for female officers was introduced.[60] The jacket had been redesigned, and rank insignia was moved from the sleeves to shoulder boards, similar to the mess dress worn by males

WORKING DRESS

Utility Jacket and Jumper

Experience with the all-seasons service dress tunic showed that it was not comfortable to work in and it did not provide adequate protection in cold weather. As an interim measure, permission was given in 1970s for non-aircrew personnel to purchase and wear the sage green flying jacket for wear in the work environment only. It was not approved for wear off the base.

The utility jacket was introduced in 1981 as the permanent all-seasons working jacket, replacing the flying jacket. Rank insignia, for warrant officers and below, was initially worn on the sleeve, but later moved to blue-grey rank slides, as used on the blue-white shirt. Medal ribbons, flying and other qualification badges were worn with this uniform but, as it was a form of working dress, medals were never worn.

---

[60] RAAF News June 1986

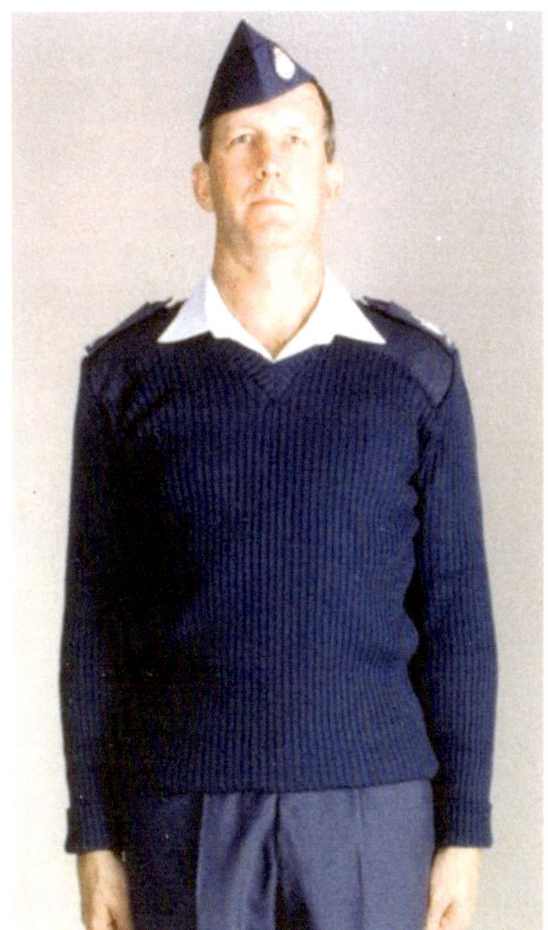

For additional protection in cold weather, a blue-grey jumper in the style of the Army's Howard Green' jumper was introduced at the same time as the utility jacket. However, the jumper was heavy and had a crew-neck which made it difficult to wear. A V-neck version replaced the earlier style, and later a lighter weight V-neck jumper was produced. A blue-grey cardigan was available for female members. On the jumpers and cardigan, the only embellishments worn were blue-grey rank slides on shoulder straps; no qualification badges were worn.

V-neck jumper

Protective Apron & Overalls

From the 1930s onwards, the working dress for those employed on strenuous or outdoor tasks such as aircraft maintenance, flight line duties and warehouse operations was dark blue, cotton overalls and a dark blue beret. The beret could be folded and placed in the airman's pocket, to be replaced on his head when he left the working area. Until about 1970, overalls were fastened in various places by plastic buttons and featured a cloth belt with a silver buckle positioned across the waistline. All new airmen/airwomen recruits were issued two sets on enlistment and the garments could be replaced free of charge when damaged or worn out. In the late 1960s, the design was changed, with zippers replacing the buttons and the removal of the belt. This was worn as a working dress until the introduction of combined working dress in 1982.

A light blue apron with dark blue pockets and trimming was worn by nurses, stores personnel and female stewards as a protective item and this was fitted with a 50 mm diameter, circular, embroidered, eagle and crown badge in light blue on blue-grey.

Light blue working dress apron

Dark blue overalls and dark blue beret without badge in the 1970s

Airwoman in combined working dress

Combined Working Dress Uniform

In 1982, overalls as general working dress were replaced with combined working dress (CWD) which was considered a smarter and more functional working dress uniform. The shirt and trousers were manufactured from a polyester/cotton twill material which was dark blue-grey in colour. The trousers were supported by the blue-grey belt that was also used in the service dress uniform. The shirt featured epaulettes on which rank slides were worn. A CWD skirt, produced for optional wear by females, was never very popular and was later withdrawn from the inventory.

All new entrants, including junior officers, received an initial free issue of two CWD shirts and trousers. Worn or damaged items were replaced by purchase from the nearest clothing store. This repayment policy was a shock to some who were used to the free exchange of worn overalls before CWD came into use.

On the shirt, a blue cloth label, with the wearer's surname embroidered in white lettering, was worn over the right

Airman (centre) wearing the tarmac jacket

pocket while a similar label featuring the letters 'RAAF' was worn over the left pocket. If the wearer belonged to a unit which had an official badge, this badge could be worn over the left breast pocket. In cold weather, the heavy duty blue jumper or the bulky vinyl tarmac jacket could be worn with the uniform. Footwear authorised for CWD were safety shoes/boots, general purpose boots, black shoes or specialist items such as firefighter boots. Personnel could choose headdress from the garrison cap, the service dress cap or an approved unit baseball-style cap.

Soon after this uniform was introduced, senior technical staff expressed concern that the exposed belt buckle could cause damage to the skin of an aircraft the person was working on. To allay these misgivings, the CWD trouser style was modified to incorporate a cloth panel enclosing the front buckle area. This new garment was only worn by those involved in aircraft maintenance; all others continued to wear the original version.

A number of initiatives occurred with CWD in the years following its introduction. In 1991, a range of CWD garments were impregnated with a fire retardant chemical. These were issued to firefighters who could then wear them as part of their working dress. Sadly, it was soon discovered that the chemical retardant washed out of the garments after five or six trips to the laundry and the impregnation process was discontinued. In 1992, CWD shorts were introduced as an optional item of wear. Two years later, the vinyl tarmac jacket, with its myriad of shortcomings, was replaced by a dark blue, polyester version, based on the jacket worn by members of the United States Navy.

The CWD garments were widely regarded as smart and functional by those who wore them. However, as a working dress, they were replaced by disruptive pattern combat uniform (DPCU).

Jungle Greens

Jungle green working dress, better known as 'jungle greens', was a cotton drill green shirt and trousers that were worn in the field in a green vegetation environment. This uniform was worn with black general purpose boots and a cotton utility hat of the same colour as the uniform.

RAAF Airfield Defence Guards in jungle green working dress. One member is in the newly issued trial pattern disruptive pattern combat uniform

When first introduced in the mid-1960s, they were only issued to airfield defence guards and ground defence officers. Other personnel undertaking weapons training or participating in field exercises wore dark blue overalls. To better equip personnel for field exercises, in the early 1970s stocks of clothing and camp equipment were held on each base for use in field activities. Jungle green uniforms and associated webbing were issued to personnel on a temporary basis and were returned on completion of the relevant activity.

In the late 1970s, jungle green uniforms were issued to new entrants at No 1 Recruit Training Unit and Officer Training School for the duration of their training. In the early 1990s, the disruptive pattern combat uniform (DPCU) was being issued to those deploying on overseas operations. At the same time, this dress replaced jungle greens as the working dress for ground defence personnel. By the mid-1990s, DPCU replaced jungle greens as the field dress for all Air Force members.

RANK INSIGNIA

Officers

On the all-seasons tunic, officers wore their rank insignia on shoulder boards, with an eagle and crown badge above rank lace, instead of rank lace on the lower sleeve.

Female officer service dress tunics were altered to include epaulettes with rank indicated by rank slides in lieu of rank lace on the sleeve. Rank slides attached to the epaulettes on the shirt were made of the same colour and material as the shirt.

Air Commodore   *Air Vice- Marshal*   *Air Marshal*   *Marshal of the RAAF*

Officers' rank slides for the light blue shirt on the same material as the shirt. 1972 to 1986:
Pilot Officer to Squadron Leader

Officers' rank slides: Wing Commander to Air Vice Marshal

Note: The most senior active rank in the RAAF is Air Marshal held by the Chief of the Air Force. On occasions when the Chief of the Air Force is appointed as Chief of the Defence Force, they hold the rank of Air Chief Marshal, with 3 standard rows of braid above the Air Rank broad row. Marshal of the RAAF has only ever been an honorary rank, most recently held by Prince Philip, Duke of Edinburgh.

Warrant Officers

In 1973, the Australian Parliament passed the Royal Style and Titles Act 1973 which required that only the Australian coat of arms, featuring the kangaroo and emu, be used by Government organisations. Up until this time, the insignia for a Warrant Officer in the RAAF and a Warrant Officer class I in the Australian Army had been the British (or Royal) coat of arms with the lion and unicorn. As this policy came into effect during the changeover period for the all-seasons uniform, the 'new' coat of arms was manufactured on the dark blue material as well as on the all-seasons blue-grey material. Anecdotal evidence suggests that serving warrant officers were reluctant to remove the traditional coat of arms badge.

On the all-seasons uniform tunic, warrant officers wore a shoulder board with only the eagle and crown badge as well as wearing the coat of arms badge on the lower part of both sleeves. On the service dress shirt, warrant officers wore a light blue rank slide with the eagle and crown badge on their epaulettes as well as the coat of arms badge on the upper sleeves of the shirt.

Warrant Officer Rank Insignia 1970 -2000

Service Dress jacket 1970-1985   *Blue 1st pattern 1972 and 2nd pattern to circa 1980*

SHADES OF BLUE

Service Dress jacket 1985-2000

Khaki Tropical Shirt 1st and 2nd patterns, 1972 to circa 1980

SD Jacket: WRAAF 1970s

Blue Shirt: WRAAF 1970s

Jungle Green: 1970s

Blue-grey Working Dress 1980s

Saxe Blue Working Dress 1980s

Blue Grey Tarmac Dress 1980s

Shirt/Utility Jacket/Safari Jacket rank slide 1985-2000   *WOFF shoulder board for Service Dress jacket*   *WOFF rank slide for shirt Withdrawn in 1980*

Warrant Officer of the Air Force

The Warrant Officer of the Air Force was a position, created in 1993, with the role of advisor to the Chief of the Air Force on matters relating to airmen and airwomen. As the person holding this position was the most senior non-commissioned person in the Air Force, a new rank insignia was created to indicate their highly respected status.

Lower sleeve of SD tunic   *Shirt epaulettes*

Warrant Officers Disciplinary

In the early 1980s, warrant officers disciplinary (WODs) were accorded the privilege of carrying a pace-stick[61] when on parades as a symbol of their position. The pace-sticks carried by unit and wing WODs were made of rosewood with brass fittings. Until the early 2000s, WODs at Air Force Headquarters and command headquarters carried ebony (black) pace-sticks with silver fittings. However, this distinction disappeared around 2005 and thereafter all WODs carried rosewood pace-sticks.

---

[61] The pace stick originated with the British Royal Regiment of Artillery ensuring the correct distance between guns. It subsequently developed into the modern pace stick to its present configuration as an aid to drill.

Senior Non-Commissioned Officers, Airmen and Airwomen

SNCOs, airmen and airwomen wore their rank insignia on both sleeves of the service dress tunic, above the elbow. On the service dress shirt, the insignia were worn on the upper part of both sleeves as well as an 'AUSTRALIA' slide on each epaulette. Drab insignia were worn on tropical dress shirts. The traditional twin-bladed airscrew insignia for a leading aircraftman (LAC) or leading aircraftwoman (LACW) was retained and the design changed to a cut-out design which no longer had a backing. Like the SNCO insignia, it was embroidered in silver/blue thread.

NCO and Leading Aircraftman/Aircraftwoman Rank Insignia 1970 – 1985

Service Dress jacket: LAC, Corporal, Sergeant & Flight Sergeant

Light blue shirt: LAC, Corporal, Sergeant & Flight Sergeant

Tropical Dress, 1970s – 1990s: LAC, Corporal (uncut), Sergeant & Flight Sergeant

Shoulder slides worn by FSGTs and below on male and female shirts and female work dresses.

Airman Aircrew Ranks 1983

The 1983 review of the airman aircrew mustering created three new aircrew ranks for qualified airman aircrew: warrant officer (air), flight sergeant (air) and sergeant (air). These ranks were equivalent in status to the non-aircrew ranks and were based on a similar system in the Royal Air Force. These airman aircrew ranks and their insignia had such a short life that some squadrons did not see them at all. They had disappeared by the end of 1983 and airmen aircrew reverted to the same rank insignia as non-aircrew personnel.

1983 Non-Commissioned Air rank insignia

1983 rank insignia for Sergeant (air), Flight Sergeant (air) & Warrant Officer (air)

Rank Insignia Changes in 1984 and 1986

In 1984, officer rank lace on the service dress tunic was moved from the shoulder board to the lower sleeve, where it had been for the first 50 years. See examples of service dress tunics earlier in this chapter.

A bigger change occurred in 1986 when the polyester/cotton service dress shirt changed to a heavier material in a colour that was described as blue-white. Rank slides for officers changed from light blue to all-seasons blue-grey colour. In addition, rank insignia for personnel from warrant officer to LAC/LACW moved from the shirt sleeve to blue-grey rank slides worn on the shirt epaulette, but they remained on the sleeve of the tunic. The 'AUSTRALIA' slide disappeared because the word was incorporated into each rank slide. Drab rank slides were produced for wear on tropical dress.

From about 1985, the rank of cadet, which was the equivalent of aircraftman, was replaced by officer cadet, which was a commissioned rank below Pilot Officer. A new rank slide featuring the white gorget was produced.

When the rank insignia of airmen and airwomen moved to the shoulder, the propeller symbol that had been used for the rank of leading aircraftman since 1921, was found to be easily confused with the single stripe of a Flying Officer. The rank insignia for a leading aircraftman/leading aircraftwoman was changed to a single chevron for both the shoulder rank slides and the sleeve insignia worn on the service dress tunic.

Officer Rank Slides – Blue-White Shirt and other uniforms – 1986 onwards

Top: Officer Cadet, Pilot Officer, Flying Officer, Flight Lieutenant, Squadron Leader.
Bottom: Wing Commander, Group Captain, Air Commodore, Air Vice-Marshal, Air Marshal

Rank Slides – Warrant Officers, Non-Commissioned Officers and Airmen/Airwomen – Blue-White Shirt and other uniforms – 1986 onwards

Top: Aircraftman/Aircraftwoman, Leading Aircraftman/Aircraftwoman and Corporal
Bottom: Sergeant, Flight Sergeant, Warrant Officer and Warrant Officer of the Air Force

Drab Tropical Dress Slides

Top: Aircraftman/Aircraftwoman, Leading Aircraftman/Aircraftwoman and Corporal
Bottom: Sergeant, Flight Sergeant, Warrant Officer and Warrant Officer of the Air Force

Cadets

With the introduction of the all-seasons uniforms, cadets both at the RAAF Academy and the Diploma Cadet Squadron (renamed Engineer Cadet Squadron in 1976), wore white rank slides on the shoulder board fitted with an eagle and crown badge (similar to a warrant officer board) with the following braid to indicate seniority within the respective establishment.

Cadet rank shoulder boards: Cadet, Air Cadet, Senior Air Cadet and Air Cadet Under Officer

On shirts and utility jackets, the same white slides shown in Chapter 5 were worn on shoulder straps.

Cadets of Engineer Cadet Squadron in all-seasons service dress uniform in 1980

Australian Defence Force Academy

With the opening of the Australian Defence Force Academy (ADFA) in 1986, RAAF cadets initially wore the same rank insignia that had been worn at RAAF Academy. When the ADFA rank structure was withdrawn in 1998, a new style of shoulder board was developed for cadets. It was worn with the newly introduced white ADFA ceremonial jacket.

Australian Defence Force Academy cadet shoulder board introduced in 1998 and the ceremonial dress uniform which was unique to ADFA

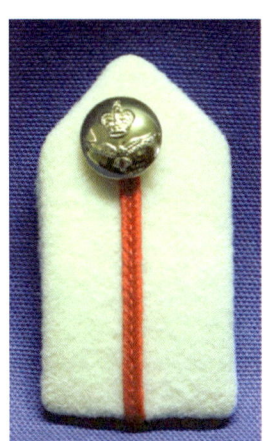

Gorget patches worn by cadets at Australian Defence Force Academy (ADFA)

Two RAAF officer cadets and an officer at the Australian Defence Force Academy in the 1990s. The cadets are wearing service dress with white hat band and gorgets while the officer is wearing Flight Lieutenant rank on his sleeve, black hat band and medical officer insignia on his collar

NCO Cadets 1983

In 1983, a new mustering of airborne electronic analyst (AEA) was created to provide non-commissioned aircrew to operate the sensor equipment on P-3 Orion maritime patrol aircraft. The plan to recruit new AEAs directly into the RAAF led to a review of the training and rank structure for all airman aircrew, including flight engineers, loadmasters, crewman technical and crewman. The review recommended that sergeant was the minimum rank for AEAs and other airman aircrew. A new rank of 'non-commissioned officer cadet' (NCO cadet) was created for trainees undergoing training to become airman aircrew. Later in the 1980s, the rank of NCO cadet was used for non-aircrew trainees who were being trained for musterings that also had the minimum rank of sergeant, such as medical laboratory technicians and environmental health staff.

Rank insignia were originally worn on the sleeve of the service dress but, when rank insignia moved to the shoulder in 1986, the NCO cadet rank slide changed to three blue embroidered outline chevrons on a white slide. A further rank slide with two embroidered chevrons was also worn during the 1980s for cadets undergoing training for musterings rated at the rank of corporal.

NCO cadet: 1983-1986 SD jacket & light blue shirt

NCO cadet shoulder slide from 1986

NCO cadet shoulder slide for Corporal musterings, 1980s

Cadet Aircrew

Course Photograph of Cadet Aircrew (CDTAC) with a Sergeant (front, centre) on the 1/84 Airman Aircrew Basic Training Course in June-July 1984. They are all wearing the All Seasons crew neck jumper of the early 1980s with shoulder badges and open neck shirts.

Apprentice Rank Slides and Insignia

Slide worn on shirt 1970 to 1985   *Worn on service dress jacket 1970 to 1980*   *New Style with merrowed edge 1980s*

When shoulder rank slides replaced sleeve insignia on shirts and utility jackets in 1986, the rank slides below were worn by apprentices to show their relative status within the training school. The apprentice warrant officer rank slide was awarded to the apprentice who was to receive the prize for 'Best Apprentice' and was only worn on the day of graduation.

Rank slides replaced sleeve insignia in the 1980s with the changeover to the blue-white shirt. They were produced with the titles "RAAF Apprentice" and "Apprentice" in all six ranks

HEADDRESS

Service Dress Cap

The all-season service dress cap was similar in design to the previous cap except that it was in the blue-grey colour. Cap badges remained the same although a cap badge featuring the Southern Cross was trialled but rejected due to manufacturing cost.

Service dress cap for an officer of air rank (Air Commodore to Air Chief Marshal)

Service dress cap for officer from Pilot Officer to Wing Commander

Service dress cap of a Warrant Officer

Service dress cap of Flight Sergeants & below, fitted with matching rainproof cover

Service dress cap of an apprentice

Service Dress Caps for Females

Working cap worn by airwomen up to and including Flight Sergeant

Service dress cap worn by airwomen up to and including Flight Sergeant. A similar style of cap was worn by officers, warrant officers and officer cadets with appropriate badge and cap band. Oak leaves were embroidered around the badge on the cap band for Group Captains

Service dress cap worn by airwomen up to and including flight sergeant. A similar cap was worn by female warrant officers but with the warrant officer badge

Service dress cap worn by female officers up to and including Wing Commander. Group Captains wore a similar cap with a single row of wattle leaves embroidered on the peak

Garrison Cap

In 1977, a blue-grey garrison cap, initially called 'forage cap'[62], was introduced as an alternative form of headdress to the bulky service dress cap. This was especially welcomed by aircrew who could carry the garrison cap in their pocket for use when away from their home base. A version for air officers (Air Commodore to Air Chief Marshal) included light blue piping on the perimeter of the cap. Garrison cap badges were made of light blue thread on blue-grey felt. Later badges incorporated gold and red colours. Officer and NCO cadets wore a white vinyl flash attached to the front of the cap while apprentices wore a light-blue vinyl flash.

Over the following years, modifications were made to both the cap and the badges. The initial badges were woven in light blue on felt. Subsequently, the position of wearing the badges moved from the bottom of the cap to the top and, about 1988, multi-coloured woven badges were introduced for the four categories of personnel.

---

[62] RAAF News, August 1975

SHADES OF BLUE

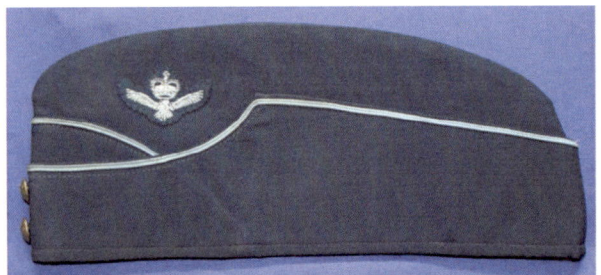
Air Officer, Maker: AGCF, Dated 1985

Air Officer, Maker: AGCF, Dated 1988

Air Officer, Maker: AGCF, Dated 1992

WOFF to GPCAPT, Maker: AGCF, 1978

PLTOFF to GPCAPT, Maker: AGCF, 1985

OFFCDT, Maker AGCF, Dated 1982

WOFF, Maker: AGCF, Dated 1988

FSGT & below, Maker: LATINERS, 1977

FSGT & below, Maker: AGCF, Dated 1977

FSGT & below, Maker: AGCF, Dated 1985

AirTC Cadet, Maker A.G.C.F., Dated 1988

AirTC Cadet, Maker ADI, Dated 1992

Cap Badges Worn on the Garrison Cap

Worn by FLTSGTs & below until 1988

WOFFs, Officers and Air Officers until 1988

Variations of the badge worn by FLTSGTs & below from 1988   *Warrant Officers*

Variation of the Officer's badge   *Variation of the Air Officer's badge*

Fur-Felt Hat

The khaki fur-felt hat continued to be worn as part of tropical dress with the traditional tri- colour flash for officers and warrant officers secured to the left side of the puggaree. Officer Cadets wore a white tape at each end of the tri-colour flash. Other ranks wore the airman's cap badge at the front.

In the early 1990s, the colour of the hat was changed to blue-grey with a blue puggaree with the appropriate embroidered garrison cap badge attached to the puggaree. The hat could be worn in lieu of service dress cap or garrison cap, subject to local commander's instructions, with all orders of dress. However, the colour of the puggaree proved unpopular, as did the identification via the cap badge, and a black puggaree with service dress cap badge was introduced. For  air rank officers, a pleat of light blue was inserted in the sixth fold of the puggaree. NCO cadets wore a white background behind the airman's badge. In the case of apprentices, the background was light-blue.

The blue-grey fur-felt hat was not very popular and was replaced by the traditional khaki hat with the advent of the Air Force Blue uniform in 2000.

Fur-felt hat with blue puggaree and airman's embroidered garrison cap badge

Fur-felt hat with black puggaree and officer's cap badge. This cap has a female cap badge, which is slightly smaller than the male badge

Beret

Berets have been part of RAAF uniforms almost from its inception, but were mostly worn by trades in conjunction with overalls and similar working dress, and without any badges for this use. The dark blue beret was worn by airmen up to the rank of flight sergeant. The beret was authorised for wear for ground defence officers and airfield defence guards in 1982, but only while wearing the jungle green working dress. Badges were the same as those authorised for wear on the garrison cap. When No 2 Airfield Defence Squadron was tasked to provide honour guards for significant RAAF events in the late 1980s, its members were authorised to wear a new blue-grey beret with service dress while performing ceremonial duties. In 1998, approval was given for the beret to be worn as a normal component of airfield defence guard uniform. It remains the only item of the all-seasons uniform still in use in 2016.

Blue-grey beret worn by airfield defence guards and ground defence officers

Chief of Air Force, Air Marshal Errol McCormack, inspecting the guard of honour in July 2001. Air Marshal McCormack was wearing the new Air Force Blue uniform. The guard consisted of personnel from No 2 Airfield Defence Squadron wearing the all-seasons ceremonial dress uniform with the blue-grey beret

QUALIFICATION BADGES: AIRCREW

When the all-seasons uniform was introduced in 1972, the three styles of flying badge in use were the same as those worn with the previous dark blue uniform:

- Bullion brevets worn on service dress jacket.
- Sterling silver brooches worn on the light blue all-seasons shirts and khaki drab shirts.
- Miniature bullion badges worn on mess dress jackets.

With the initial all-seasons uniform, there was no equivalent of the battle dress blouse and so the need for embroidered flying badges had disappeared. However, when the utility jacket was brought into use in 1981, embroidered flying badges were re-introduced.

In 1972, the flying badges being awarded were pilot, navigator (N), air electronics officer (AE), flight engineer (E), loadmaster (L) and crewman (C). In addition, the following flying badges were still being worn by previously qualified members, but no new members qualified for them:

- The signaller (S) brevet, worn by those who had not re-categorised to air electronics officers.
- The assistant crewman or gunner (G) brevet, worn by gunners trained for Iroquois helicopters operating in Vietnam. This role ceased in 1971.
- The parachute jump instructor brevet worn by PJIs qualified before transfer of the role to the Army in 1972.

Over the period 1972 to 2000, the following new aircrew brevets were introduced by the Air Force.

Embroidered Crewman Technical 1977-1983

Embroidered Flight Steward 1975-2002

Sterling silver Flight Technician 1995-2004

Sterling silver Fighter Controller 1996-2007

Officer Aircrew non-pilot 'Southern Cross' brevet

Airborne Electronic Analyst 1984
(Re-use of previous brevet)

New blue-grey backing for bullion aircrew flying badges introduced briefly in mid 1990s

Crewman Technical 1977-1983

Non-commissioned aircrew drawn from technical musterings were authorised to carry out minor maintenance and repairs on aircraft away from their home base, but aircrew drawn from non-technical trades were not. To identify those aircrew with technical skills, the new aircrew mustering of crewman technical was created in 1974 and a crewman technical brevet was introduced in 1977. It was worn by technically qualified aircrew on Iroquois, Chinook, Dakota and Caribou aircraft.

In 1983, crewmen technical were reclassified as flight engineers and the CT brevet was no longer awarded. Some crewman technical brevets were manufactured with the monogram letters reversed to TC, but these were never issued. (See Appendix 10)

Flight Steward 1975-2002

Although flight stewards had been flying on Air Force VIP aircraft since the early 1950s, no flight steward brevet was in use until the first were awarded in 1975. This was the first brevet that female members could qualify for.[63] The FS brevet was replaced by the crew attendant (CA) brevet in 2002.

Flight Technician 1995-2004

When RAAF HS748 transport aircraft were operating away from home base, maintenance personnel were carried, when needed, to perform maintenance and assist with loading. From 1995, the role of these members was recognised by the award of the flight technician (FT) brevet. When the HS748 aircraft were withdrawn from service in 2004, the FT brevet was no longer awarded.

Fighter Controller 1996-2007

Until the early 1990s, Air Force air defence officers were ground-based and were not entitled to wear a brevet. In the mid-1990s, those air defence officers who had served a tour flying as controllers on airborne-early-warning-and-control (AEW&C) aircraft with the RAF, US Navy or US Air Force were recognised as aircrew and awarded the fighter controller (FC) brevet which was based on a similar one used by the RAF. The first FC brevets were awarded in 1996. This brevet was superseded in 2007 when air defence officers were re-categorised as air combat officers and were entitled to wear the officer aircrew non-pilot brevet, also known as the 'Southern Cross' brevet. Fighter controllers and air defence officers later became aircrew on the Wedgetail AEW&C aircraft when it entered RAAF service in 2012.

'Southern Cross' Brevet 1998

In 1998, the navigator (N) brevet was replaced with the officer aircrew non-pilot brevet which was similar to the two-winged pilot brevet but with the RAAF monogram replaced by a stylised Southern Cross constellation. The first of these brevets was awarded in December 1998. Navigators who had graduated before this date continued to wear their navigator brevet. From 2006, this brevet was worn by all members of the air combat officer category which included navigators, air electronics officers and air defence officers.

Airborne Electronic Analyst 1984

When the training of air electronics officers ceased in 1979, navigators initially took over the sensor positions in the P-3 Orion aircraft. Eventually, a new aircrew mustering of airborne electronic analyst was created in 1983 and its members initially wore the rank of sergeant. When the first airborne electronic analysts completed their training in July 1984, they were awarded the AE brevet. This brevet had therefore been in

---

[63] RAAF News, July 1975 edition, Canberra ACT, p 3

continuous use since 1965, but the qualification had changed.

New Background Colour

For decades, the backing for both embroidered and bullion aircrew brevets was a black material and it remained in place even when the blue-grey all-seasons uniform came into use in 1972. In the mid-1990s, the backing material was changed to blue-grey to provide a better colour match with the garments on which these badges were worn. Not long after this change was implemented, the Air Force Blue uniform arrived requiring the backing material to be changed back to black again.

QUALIFICATION BADGES: GROUND STAFF

Medical Branch Badge 1984-1986

When the all-season uniform was introduced and officers' rank insignia was located on shoulder boards, medical branch officers wore the Rod of Aesculapius badge on their shoulder boards. Non-commissioned members of the medical branch wore the badge on each lapel of the tunic. If the tunic was not worn, the badge was worn on each collar of the shirt. When shoulder boards disappeared in 1984, medical officers also wore the badge on each lapel and it was worn on the rank slides of all medical branch personnel after 1986. The illustrated shoulder board for a Flight Lieutenant in the Medical Branch, shows the new Medical Branch badge which replaced the eagle and crown badge.

During the 1970s and 1980s, two variants of the medical branch badge were issued but, by the 1990s, the medical branch had reverted to the style of badge used in the 1960s.

Aero-Medical Evacuation Surgeon 1982-1990

The badge was manufactured about 1982 and remained on the clothing list (NSN 8455-113-4471) until about 1990. Colloquially known as "Florence's Lamp", it has not been possible to find any photographic evidence of it being worn.

Medical Branch 1984-1986   *Aero-Medical Evacuation Surgeon*

Skill-at-Arms Badge 1985 -2002

From 1985, the skill-at-arms badge was worn on the left sleeve of the service dress tunic, the blue-white shirt and the drab shirt. No official mess dress version was produced but some personnel were known to wear the white Army version of this badge on the mess dress jacket.

Parachutist 1985

A badge to recognise those members who were qualified in military parachuting was introduced in November 1985. The design was an adaptation of the Army parachutist badge but in a colour scheme compatible with the all-seasons blue-grey uniform.[64] Initially, the badge was approved in three forms: the blue-grey woven cloth version for wear on all items of service dress, a gold bullion miniature for wear on mess dress and a blue wing on khaki background for wear on tropical dress. In 1997, an embroidered version replaced the woven cloth badge. The parachutist badge is worn on the right sleeve of the service dress tunic, the blue-white shirt or the mess dress jacket.

Parachutist badge worn from 1985 to 1996   *Parachutist badge worn from 1997 to 2000*

Legal Officer Collar Badge 1989

This was introduced in 1989 and is worn by Air Force Legal Officers. They work as part of the Australian Defence Force joint legal team and can be posted to single service and joint units to advise on the complete range of legal issues within the ADF.

---

[64] John O'Connor, Australian Airborne: The History and Insignia of Australian Military Parachuting, John O'Connor, Kingsgrove, NSW, 2006 pp 144-147

RAAF Police Badges 1975 & 1996

In 1975, the RAAF police replaced the familiar 'SP' brassard with a chest or fob medallion, with individual numbers, worn on the left breast pocket of service dress outer garments. Officer's badges were gold in colour and silver for other personnel. The number '007' was not allocated. The design of the badge was based on the griffin of Greek mythology, which has a lion's body and an eagle's head and wings. Both badges were replaced by a fob for all ranks in 1996.

Officer's Fob *Other Rank's Fob* *Dog Handler* *1996 All Ranks*

OTHER BADGES

Australia Badges

The 'AUSTRALIA' shoulder badge, previously only worn by members serving outside Australia, became standard on the all-seasons tunic. For a short period in the late 1970s, the shoulder badge changed to 'ROYAL AUSTRALIAN AIR FORCE' before the 'AUSTRALIA' badge returned.

Shoulder Devices

The royal cypher shoulder device was worn by an officer holding an honorary appointment to the sovereign, such as aide-de-camp or equerry. With the all-seasons tunic, the shoulder device was worn on the shoulder board superimposed on the rank braid. When officer rank insignia moved to the sleeve in 1984, the shoulder device was still worn on the shoulder, fixed in place by cloth strips the same colour as the tunic. When no tunic was worn, the device was worn on the rank slide superimposed on the rank braid.

Shoulder board for a group captain with the cypher of Queen Elizabeth II

ACCOUTREMENTS

Brassards

These were traditionally reserved for members actually carrying out a specified duty and now comprised:

Orderly Officer.

This was made from black cloth with a scarlet edge and the letters 'OO' in scarlet cloth. This brassard was superseded by the orderly member brassard in the early 2000s.

Orderly Non-Commissioned Officer (not pictured). This brassard was made from black cloth with scarlet edge. It was withdrawn from use in the early 2000s and replaced by the Orderly Member brassard with letters 'OM'.

Air Movements. The brassard was made from red cloth, 9cm wide with an eight-spoked yellow wheel of 7.6cm diameter.

Bomb Disposal. This brassard was made of light blue cloth, 9cm wide, with a bomb separating the letters 'B' and 'D' within a laurel wreath embroidered in light blue on a dark blue background.

Security and Policing Service Brassard. This brassard was of black cloth with the letters 'SP' centrally embroidered in white lettering. Before 1975, it was worn on the upper right arm attached to the epaulette by all members of the RAAF Police service when on duty. After the police fob badge was introduced in 1975, the brassard was only worn with the following uniforms:

- Combined working dress
- Disruptive pattern combat uniform (DPCU)
- Service dress when the jumper was worn

Health Services Brassard. The health services brassard was based on DPCU material with a red cross embroidered on a white background. The brassard was worn by all health services personnel and chaplains deployed on exercises or operations, to indicate their non-combatant status.

General Service Instructor. This brassard was worn by general service instructors who trained new entrants to the RAAF at initial training units. It contained the general service instructor insignia consisting of an open pace-stick surmounted by a horizontal F88 Steyr rifle and an eagle. The pace-stick represented drill while the rifle represented ground defence—the two major roles of the general service instructor. The eagle represented the Air Force environment that the training was conducted in. This brassard was introduced in 1996 and was superseded by the military skills instructor brassard in 2001. Initially, the brassard was just the letters 'GSI' without the insignia.

TRIAL PATTERNS

During the period of the all-seasons uniform, many trials were undertaken both of uniforms and badges. Some were adopted but many were not. One trial design that stands out as being a major departure from tradition was the redesign of the pilot brevet. The reasons for the design being trialled and rejected are lost but it is an interesting historical point that the current pilot brevet, which has largely been in use unaltered for nearly a century, could have been quite different.

Bullion pilot brevet proposed in 1979

Sterling silver pilot brevet proposed in 1979

There were no changes to cap badges, although a metal officer badge featuring the "Southern Cross" was trialled but rejected due to manufacturing cost.

Southern Cross officer's trial cap badge

End of the All–Seasons Uniform

Although the introduction of the all-season uniform went very smoothly, there was one flaw in the process. When the contract with the material manufacturer was drawn up, there was no clause added to prevent the supplier from selling the new blue-grey material to other organisations. As a result, many government agencies and private companies acquired the blue-grey cloth for their own uniforms and the 'uniqueness' of the new Air Force uniform was somewhat lost. Because of this, there were many reported cases of RAAF personnel being mistaken for railways or customs staff. As the end of the century approached, this problem contributed to the change of colour back to Air Force Blue, which is covered in the next chapter.

Engineer Cadet Squadron 22 Course 1980, Cadet Under Officers 1981, at RAAF Frognall VIC. The sole female Cadet is Jenny Fantini (now Group Captain)

CHAPTER 7:
THE RAAF RETURNS TO DARK BLUE—2000 to 2016

The All Seasons (blue-grey) uniforms had been in use since 1972 and, from the time of its introduction, there had been a degree of dissatisfaction amongst RAAF personnel with the lack of service identification it provided. Many considered that the basic problem lay with the considerable number of government organisations and private companies who were able to use the same colour and material in their own corporate clothing. This led to Air Force personnel being confused with staff from other companies and agencies. Many personnel also felt that the style of the uniform did not look 'military'.

Consequently in 1998, the Chief of Air Force (CAF), Air Marshal Errol McCormack, decided that the time was opportune to change the uniform to one which would readily identify the wearer's parent service and project a more military image. The 'All Seasons' concept was to be retained by having a range of garments that would make the uniform practical and comfortable to wear in climates ranging from winter in a temperate zone to tropical. This concept included the principle that the uniform worn on a particular activity would be decided by the local commander, not by the calendar.

CAF directed that the change process should involve four main areas—colour, style, embellishments and accoutrements. This was how it was implemented.

Colour

A change of colour was seen as an essential element in both promoting wearer pride in the Air Force uniform and establishing a unique image. The colour of the All Seasons uniform was regarded by many as 'insipid' and 'non-military'. It was felt that the only way to dispel this RAAF-wide negative perception was by the choice of a new colour, far removed from the light blue-grey then in widespread use by government agencies and the private sector.

After examining the colours in use by other air forces, the decision was made to revert to dark blue, the same colour that had been used for the first half century of the RAAF's history. The following were the principal reasons for this choice:

- At the time, very few organisations used dark blue in their uniforms, except for some police forces.
- Dark blue was the traditional colour of the Air Force, having been the colour of the uniforms from 1921 to 1972.
- A dark blue uniform, with appropriate embellishments, presented a smart and military image on ceremonial occasions.

Clothing technologists identified this shade on their database as Number 95903377 and it was designated by the RAAF as Air Force Blue.

Style

The negative comments regarding the style of the All Seasons uniform included 'looking like a lounge suit' and 'based on the fashions of the 1970s'. The existing style of jacket had side vents rather than a single vent, no side pockets and a draped cut rather than a waisted one. From a practical viewpoint, this style had a number of drawbacks. The side vents accentuated any rear protuberances, the tight cut made any movement difficult and the draped pattern meant accoutrements such as the ceremonial sash could not be worn in a smart manner.

In light of these considerations, a change to the style of the 1A uniform was considered warranted, both from a practical and aesthetic point of view. The decision was made by CAF to revert to the pre-1972 style of jacket. Advantages of this style were seen as the following:

- Patterns for this style of garment were still in existence.
- The jacket had a military look with its belt and side pockets.
- No large organisation outside the Australian Defence Force had a uniform of similar style and, because of its military cut, imitations were less likely to occur in the future.
- The waisted cut of the pre-1972 uniform permitted the wearing of accoutrements in a much smarter and more comfortable fashion.
- The jacket had a single vent, which provided more comfort and a better appearance for the wearer.

Material

The material for the new Air Force blue uniform was 75 per cent wool and 25 per cent polyester. In later years, a lightweight uniform was introduced for wear in warmer climates. The material for this uniform was 50 per cent wool and 50 per cent polyester. The first person to wear this garment on a trial basis was Group Captain Leo Davies, now the current Chief of Air Force. He was very impressed with the uniform and his positive report led to its adoption into the RAAF order of dress.

A contract for supply of the material was awarded to Macquarie Textiles, a firm based in Albury NSW. One important element of this contract was that the material was not to be supplied to any other organisation or individual. This strict clause was inserted because the manufacturer of the previously used blue-grey material had made it widely available to other organisations, thus creating an identity problem for Air Force personnel.

Introduction

From 2000, all RAAF personnel received an initial free issue of Air Force Blue trousers/slacks, tie, belt, cap and black socks. All new garments were manufactured by the firm Australian Defence Apparel (ADA), based in Melbourne. The service dress 1A uniform (tunic with trousers or skirt) was supplied on a 'made to fit' basis. This involved an individual attending the local clothing store where his/her measurements were taken, which were then forwarded to ADA. Within six weeks a 'try on' uniform was provided and adjustments made by local tailors or seamstresses, if possible. If major adjustments were needed, the uniform was returned to ADA for further changes.

Service Dress

Service dress remained the dress that was worn by members representing the RAAF on day-to-day business with the Australian public or overseas.[65] It continued to be the uniform most often seen by the public.

Service dress 1A consisted of the Air Force Blue tunic with gilt buttons, gilt belt buckle and ribbons. Rank insignia was worn on the sleeve, with officer rank braid and warrant officer insignia worn on the lower sleeve while non-commissioned officer and airman rank was worn on the upper sleeve. The tunic was worn over a blue-white shirt with a tie, and with Air Force Blue trousers and black leather shoes. Other versions of service dress allowed either a long-sleeved shirt with tie or a short-sleeved shirt to be worn with trousers but no tunic. On the shirt, rank insignia was always worn on epaulettes. Female members could wear either an Air Force Blue skirt or slacks.

When the uniform changed from the All Seasons blue-grey to Air Force Blue, the colour of the tie and belt also changed from blue-grey to Air Force Blue. Socks changed from blue-grey to black—the colour they had been prior to the introduction of the All-Seasons uniform. The blue-white shirt that had been worn with the All-Seasons uniform since the 1980s in both long-sleeved and short-sleeved versions continued to be worn.

---

[65] Note that much of this chapter is based on AAP 5135.003 RAAF Manual of Dress dated 30 July 2015.

Different versions of service dress. Another version allowed the short-sleeved shirt without a tie

An alternative service dress for officers and warrant officers, particularly in tropical areas, was the Air Force Blue safari suit. It could be worn as a working dress, as service dress for a unit parade or as ceremonial dress. In the latter situation, it was worn with sword and medals.

Patent leather shoes were originally introduced as an optional item of wear in lieu of the plain leather shoes. Subsequently they became standard footwear on ceremonial occasions.

Wing Commander Robert Graham inspecting an Australian Air Force Cadet unit in Darwin about 2010. He is wearing a safari suit with medals

Ceremonial Dress

Ceremonial dress was based on service dress with the addition of medals and arms—swords for officers and warrant officers or rifles for flight sergeants and below. Other accoutrements such as blue shoulder sash, aiguillettes and white ceremonial belt may also be worn with this uniform depending on the occasion.

Chief of Air Force, Air Marshal Geoff Brown AO inspects a parade at RAAF Base Amberley on 4 July 2013. Air Marshal Brown was wearing ceremonial dress for officers for Air Vice-Marshal and above. It consisted of service dress tunic with medals, gold sash with sword and an aiguillette on the right shoulder. Other members present are also wearing ceremonial dress consisting of service dress with medals and arms, either sword or rifle

Tropical Dress

The khaki tropical dress, (Chapter 6) better known as 'drabs', was abolished in 2013. For the first time since the RAAF was formed in 1921, the service did not have a khaki uniform in its wardrobe.

Mess Dress

No changes were made to mess dress with the introduction of the new Air Force Blue uniform in 2000. Uniforms for officers and trousers/skirts for warrant officers and senior non-commissioned officers continued to be manufactured in dark blue material.

In 2015, major changes were made to mess dress. The white mess dress jacket worn by officers in summer and warrant officers and senior non-commissioned officers all year was withdrawn. For males, it was replaced by an Air Force Blue jacket and trousers that were worn in both summer and winter. All rank insignia was worn on redesigned shoulder boards which are shown in the later section on rank insignia. An Air Force Blue cummerbund was also worn.

On formal occasions, females continued to wear the mess dress that they had worn since the 1990s. A suitable female mess dress was still being developed in 2017.

Air Force blue mess dress for males. Females remained in the 1990s mess dress

For hot climates, the new mess dress included a hot weather version which could be worn in tropical areas or in temperate areas in summer. This version was commonly known as 'Red Sea rig'.

WORKING DRESS

Tarmac Dress

Tarmac dress was a practical working dress for aircraft maintenance technicians working on aircraft on the tarmac (as the name implies) or in the hangar. This dress was worn by maintenance personnel of flight sergeant rank and below and was only worn on base in working areas. It was commonly worn with a cap bearing the squadron colours and badge. A cold weather jacket could be worn during cooler periods.

Little has changed in tarmac dress over the years

Disruptive Pattern Combat Uniform

As described in chapter 6, the standard working dress for airmen away from the tarmac was the dark blue combined working dress (CWD). In 2004, CWD was replaced by the disruptive pattern combat uniform (DPCU) which was based on the Army DPCU but with RAAF insignia. This change was introduced because, during this period the Air Force was involved in a number of deployments for which DPCU was the required dress. Members also appreciated the minimum care required for DPCU garments compared with the higher standards applicable to CWD items. Another benefit was that Air Force personnel were required to replace CWD garments by purchase through clothing stores, whereas their Army counterparts were entitled to unlimited free exchange of DPCU clothing.

Security Police, Military working dog handler dressed in disruptive pattern combat uniform

DPCU uniform could be worn with the DPCU utility hat or the khaki fur-felt hat. A disruptive pattern desert uniform was also worn on operational deployments to the Middle East.

Corporal wearing desert pattern combat uniform

Chief of Air Force, Air Marshal Geoff Brown AO, and Warrant Officer of the Air Force, Mark Pentreath CSM, in the new Air Force general purpose uniform

General Purpose Uniform (GPU)

One problem of using DPCU as a working dress was that, on many occasions, Air Force personnel wearing DPCU were mistaken for Army members. In 2014, an Air Force unique camouflage uniform known as general purpose uniform (GPU) was introduced as a working dress to be worn instead of DPCU in most field situations. DPCU was retained for wear in situations where camouflage was essential. A new, dark blue, cold weather jacket to be worn with GPU was introduced in 2016.

Other Working Dress Items

The utility jacket that had been introduced in 1981 in blue-grey colour (Chapter 6), was re-issued in 2000 in Air Force Blue. It was worn with embroidered qualification badges and with rank insignia on rank slides on the shoulder. This jacket was later withdrawn from use. A leather jacket (illustrated) was introduced for wear as a working dress outer garment in cold weather. It was worn over the long sleeved blue-white shirt and Air Force Blue trousers. Attached to the jacket were Air Force Blue rank slides on the shoulder, a name patch on the left breast and a unit badge or Air Force badge on the right breast. The leather jacket could be worn with or without a tie.

Another significant innovation in both working conditions and the new range of uniforms, was the inclusion of a maternity dress for wear by females after the first trimester of their pregnancy. Several versions were available.

RANK INSIGNIA

Service dress

Maternity uniform

On the new uniform, the rank insignia did not change; they were simply put on Air Force Blue backing material. On the service dress tunic, officer and warrant officer rank insignia stayed on the lower sleeve, while for other ranks, the rank badges remained on the upper sleeve.

On the service dress shirt, rank insignia were sewn on slides that were placed over shoulder epaulettes for all ranks. The officer slides used the same type of composite rank lace as used on the lower sleeve of the service dress tunic. These rank slides were also worn on the general purpose uniform.

Air Force Blue rank slides for Flight Sergeant, Group Captain, NCO Cadet and Officer Cadet

Mess Dress

From July 2006, the rank boards worn with mess dress incorporated 'AUSTRALIA' embroidered in gold at the outer end of the board. The rank was shown in gold lace for officers or by gold embroidery for warrant officers and senior non-commissioned officers, surmounted with a gold eagle-and-crown badge. With the eagle badge on the shoulder board, flight sergeants and sergeants no longer wore the eagle badge on the lapels of the jacket.

HEADDRESS

Service Dress Cap

With the new uniform, the style and badges of the service dress cap remained the same but the colour changed from blue-grey to Air Force Blue.

Service Dress cap: Airman/Airwoman to Flight Sergeant

Service Dress cap: Warrant Officer

Service Dress cap: Pilot Officer to Wing Commander

Service Dress cap: Group Captain

Service Dress cap: Air Rank Officers

Service Dress Caps for Females

When the new uniform was introduced in 2000, females wore the same Air Force Blue service dress caps as males. In 2009, a different style of cap was introduced for females. The badges on the female officer caps were the same design as those on male caps, but slightly smaller.

Khaki Fur-Felt Hat

The blue-grey fur-felt hat that was issued in the 1990s was replaced by the familiar khaki fur-felt hat fitted with an Air Force Blue puggaree and a service dress cap badge according to the wearer's rank. This hat could be worn with most forms of dress particularly during hot/tropical weather

Khaki fur-felt hat with officer's cap badge

Garrison Cap

Like the service dress cap, the garrison cap remained the same but changed to Air Force Blue colour.

Garrison cap: FLTSGT & below

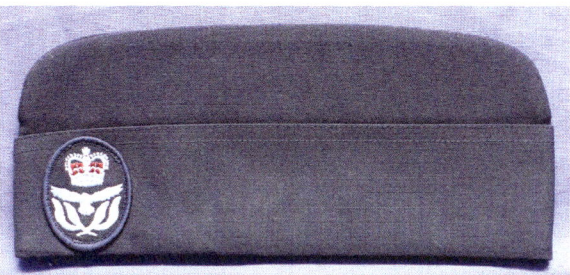

Garrison cap: Warrant Officer

Garrison cap: Pilot Officer to Group Captain

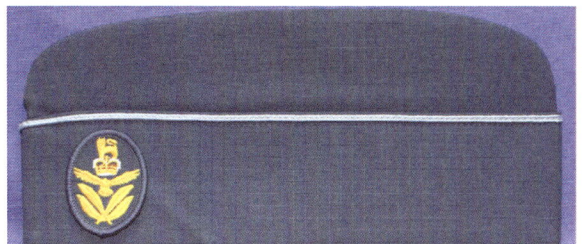

Garrison cap: Air Rank Officers

Beret

Despite the change in uniform colour, the colour of the beret worn by airfield defence guards and ground defence officers remained the blue-grey colour. In mid-2013, RAAF security musterings and personnel were re-organised into new security force squadrons. In addition to airfield defence guards, the new squadrons included members of the air force security and air force police musterings. The privilege of wearing the blue-grey beret was extended to all members of these musterings.

Air Commodore John Meier presenting an award to an airfield defence guard

In recognition of the special role undertaken by combat controllers, they were authorised to wear a unique form of headdress—a dark grey beret with the combat controller badge when wearing working dress.

Dark grey beret with combat controller badge and black backing shield

QUALIFICATION BADGES: AIRCREW

With the change to Air Force Blue, the existing range of flying badges was manufactured in the same design and material but with the background colour now changed to Air Force Blue. In 2000, flying badges in use were pilot, non-pilot officer aircrew (see following section 'Southern Cross' brevet), flight engineer (E), loadmaster (L), airborne electronic analyst (AE), fighter controller (FC), flight steward (FS) and flight technician (FT).

Embroidered flying badges on Air Force Blue background

In 2000, the qualification of the cabin crew on Air Force Boeing 707 and VIP aircraft changed from 'flight steward' to 'crew attendant' to recognise additional responsibilities being placed on these crewmembers. Reflecting this, the FS brevet was changed to a CA brevet in 2002. Members trained to the crew attendant standard changed to the CA brevet but those no longer employed in this role continued to wear the FS brevet.

With the retirement of the Hawker Siddeley 748 transport and navigation trainer aircraft from service in 2004, the requirement for the flight technician role ended and the FT brevet was no longer issued. Personnel who had qualified as flight technicians continued to wear the brevet for the remainder of their career.

Sterling silver badges remained unchanged apart from some changes to aircrew categories.

'Southern Cross' Brevet
(Non-Pilot Aircrew 1998, 'Air Combat Officer' 2006, 'Mission Aircrew' 2018)

The biggest change in flying badges came with the 'Southern Cross' brevet. This brevet, officially known as 'Non-Pilot Officer Aircrew', was awarded to all navigators who graduated after 1 December 1998. Navigators, air electronics officers and fighter controllers who had qualified in their category prior to this date were permitted to wear the single-wing brevets that they had been presented with. In 2006, these three aircrew categories and the category of air defence officer were merged to form the air combat officer category. All members of this new category wore the 'Southern Cross' brevet, including those who had qualified as aircrew before this brevet had been introduced. The category was renamed again in 2018 as Mission Aircrew. The brevet came in four versions—embroidered in white silk on Air Force Blue for the utility jacket, full size bullion for service and ceremonial dress, miniature bullion for mess dress, and sterling silver for wearing on a shirt.

Non-Pilot Aircrew 1998, 'Air Combat Officer' 2006, 'Mission Aircrew' 2018 brevet

Embroidered, silver brooch, full size bullion, and miniature bullion forms

QUALIFICATION BADGES: GROUND STAFF

Parachutist Badge

The parachutist badge has been worn by those members qualified to do military parachuting and was previously based on the Army's parachutist badge. As part of the introduction of the Air Force Blue uniform, the badge was changed from the irregular frame of the Army version to an elliptical shape of Air Force Blue cloth. This badge was also made in a khaki version for tropical dress and a miniature bullion version on white background for wear on white mess dress jackets. The badge is worn on the right sleeve of the service dress tunic, the blue-white shirt or the mess dress jacket.

Parachutist badge for service dress tunic and blue-white shirt

Miniature bullion parachutist badge for wear on mess dress

Skill at Arms Badge

The skill at arms badge recognised those who had qualified as a marksman using the Steyr rifle, after 1 August 1997. It was available in a number of versions including a khaki badge for wear on tropical dress. However as of 2017, a miniature bullion version for wear on Air Force Blue mess dress jackets had not yet been developed. The badge is worn on the left sleeve of the service dress tunic, the blue-white shirt or the mess dress jacket.

  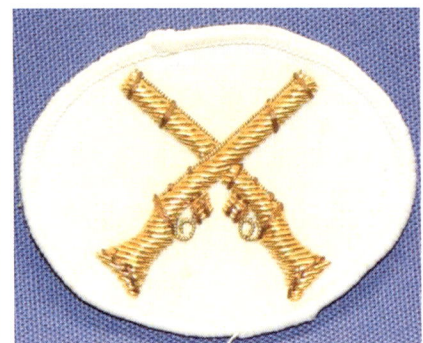

Skill at arms badges for service dress tunic, blue-white shirt and white mess dress

Military Skills Instructor

The military skills instructor (MSI) badge was introduced in 2003 to recognise the skills of instructors who teach drill, weapon handling and other military skills to recruits at Air Force training schools. These staff were previously identified by a brassard. The badge was worn on the flap of the right breast pocket of the service dress tunic or the blue-white shirt.

Initially, the badge could only be worn whilst instructing at these units and was to be removed on posting. However, in 2010, regulations changed to allow the badge to be worn by qualified personnel for the remainder of their Air Force service.

Air Force Ground Combat Badge, Infantry Combat Badge and Army Combat Badge

In November 2007, approval was given to wear the Australian Army's infantry combat badge or Army combat badge on RAAF uniforms for any member who had qualified for either of these badges during previous service. The infantry combat badge is awarded for service as an infantryman in warlike operations. The Army combat badge recognises the unique service of a member operating with an Army corps unit within a warlike area of operations. Both of these badges are made from bronze material.

In 2013 the Air Force ground combat badge was introduced to recognise service by Air Force members in a high-threat environment during deployment to a warlike area of operations. Eligibility for the badge is backdated to the Vietnam conflict. The badge has a pewter finish.

These badges were worn on the left breast of the service dress tunic or the blue-white shirt above any ribbons or medals. A miniature version was worn on the mess dress jacket.

Infantry combat badge   *Army combat badge*   *Air Force ground combat badge*

Executive Warrant Officer

In 2008, the RAAF introduced the new mustering of executive warrant officer who is the senior enlisted member within their headquarters who directly assists commanders to address strategic issues. These highly regarded members are identified by the executive warrant officer badge which was worn on the left breast above any ribbons, medals or brevets. On mess dress, the miniature executive officer badge is worn on the left lapel above any miniature medals.

Executive Warrant Officer badge, miniature and full size

Warrant Officer of the Air Force Mark Pentreath, who held this appointment from 2011 to 2015. He is wearing the Army Combat badge and the Executive Warrant Officer badge above his medals

Combat Controller Badge and RAAF Commando Badge[66]

After several years of development, the mustering of combat controller was introduced in 2009. Combat controllers give guidance to combat aircraft engaged in close air support of ground units, often designating enemy positions and targets from the ground. They also do reconnaissance of tactical landing zones ahead of the insertion of ground forces by fixed-wing aircraft. Combat controllers work closely with Army Special Forces and commando units, so part of their training is to complete commando training. Those qualified wear the RAAF commando badge which is based on the equivalent Army badge.

RAAF commando badge

In recognition of the special role undertaken by combat controllers, they wear a unique form of headdress—a dark grey beret with the combat controller badge. The badge is manufactured in pewter coloured metal and is made up of the following elements:

---

[66] The contribution of Wing Commander Harvey Reynolds, Commanding Officer No 4 Squadron, RAAF Williamtown in preparing this section is acknowledged.

- The commando dagger signifies the close ties with Army commando units.
- The shield signifies the protection of ground forces, civilians and aircrew from harm in combat.
- The wings represent the airpower integration role of combat controllers.

Embroidered Combat Controller badge for wear on the utility jacket

Combat Controller badge worn on the dark grey beret

Air Force Police

In 2013, a new version of the fob badge worn by Air Force Police was issued. The badge was buttoned to the left breast pocket. The member's police number is inscribed on the lower panel of the badge.

Joint Battlefield Airspace Controller

Joint battlefield airspace controller sterling silver brooch

In 2009, the RAAF air traffic control officer category was renamed joint battlefield airspace controller (JBAC) to better reflect the true nature of military air traffic control in modern joint operations. In early 2010, the Chief of Air Force formally approved the JBAC badge design which incorporated the stars of the Southern Cross displayed in a vertical manner as on the Australian National Ensign rather than the slanted device on the RAAF Ensign. Full size bullion, miniature bullion and cloth versions were approved later that year and the first badges were awarded in September 2010. The badge is equivalent to an aircrew flying badge and is worn in the same manner.

Joint battlefield airspace controller bullion (left and centre) and embroidered (right) badges

OTHER BADGES

A number of new badges were introduced as part of the implementation of the Air Force Blue uniform. Principal amongst these were the following:

Individual Readiness Badge

This badge was introduced as the operational readiness badge on 27 April 2000 but the name changed to 'individual readiness badge' on 18 August 2005. The badge comprises an eagle surmounted on a laurel wreath. It is worn on the flap of the right breast pocket of various garments by those who are qualified to deploy on operations. Until 2013, a gold-plated version was worn by members holding attaché posts in Australian embassies and high commissions around the world.

Individual readiness badge in antique silver and the gold-plated version

The Air Force Patch

The embroidered Air Force patch, colloquially known as 'the biscuit', is an oval-shaped badge which encloses an eagle over the words 'Air Force'. Its purpose is to more clearly identity the wearer as a member of the Air Force. It is worn on the upper left sleeve of blue-white shirts, pullovers and utility jackets. The most common form is white lettering on Air Force Blue background, but khaki and white versions are also available for wear on other uniforms.

When disruptive pattern combat uniform (DPCU) and disruptive pattern desert uniform (DPDU) were issued in the early 2000s, first for combat operations in the Middle East and later as a working dress, an Air Force patch in camouflage colours was used to identify members. In 2010, the design of these badges was changed to the RAAF roundel containing the kangaroo symbol.

DPCU (Combat) & DPDU (Desert) Air Force patches up to 2010

DPCU (Combat) & DPDU (Desert) Air Force patches from 2010

Brassards

Brassards (armlets) were provided to enable ready identification of personnel undertaking specific duties. They are worn on the upper left arm only, with the exception of the Service Police brassard, which is worn on the upper right arm. All brassards, with the exception of the SP Brassard are to feature the appropriately coloured 'Air Force' patch at the top of the brassard in a position similar to it being worn on the garment.

In 2014, the following brassards were being worn:

Orderly Member brassard is of black cloth 9.5 cm wide, with 4 mm scarlet edge, and the letters 'OM' in scarlet centrally placed.

Movements brassard is made of red cloth, 9 cm wide with an 8 spoked yellow wheel centrally placed.

Bomb Disposal brassard is of Air Force Blue cloth, 9 cm wide, with a bomb symbol separating letters 'B' and 'D' within a laurel wreath embroidered in light blue. A version on camouflaged background is also available for wear with DPCU or DPDU.

Military Skills Instructor brassard is of the same colour material (Air Force Blue or camouflage) as the uniform being worn at the time. The letters 'MSI' are surrounded by a laurel wreath stitched in a colour that contrasts with the background.

Service Police brassard is made of black cloth with the letters 'SP' centrally embroidered in white lettering. The brassard is only worn by officers and airmen of the RAAF Security and Policing Service and is worn on the upper right arm attached to the epaulette. There are also RAAF Security Police versions for appropriate uniforms.

Red Cross brassard is made of DPCU material with a red cross embroidered on a white square background. The DPCU 'Air Force' badge is located centrally 1 cm below the top seam. This brassard was worn by all health personnel, chaplains and their staff deployed on exercises or operations.

Red Cross DPCU brassard   *Reserve Airbase Protection Flight member on duty at Avalon Airshow in 2015*

Reserve Airbase Protection Flight brassard is made of black cloth with the wording 'RAAF SECURITY' embroidered in white lettering. This brassard is worn on the upper right arm by members of the Reserve Airbase Protection Flight when on security duty.

Generic RAAF DPDU Brassard During operations in the Middle East in the early 2000s, a generic desert camouflage brassard came into use to which various badges could be attached to show the duty of the wearer. This desert camouflage brassard, with the Australian National Flag badge, was used to identify wearer as an RAAF member.

Generic RAAF Air Force Blue Brassard

In 2015, with the introduction of the 'blue camouflage' general purpose uniform, an Air Force Blue brassard came into use in Australia. This was fitted with a Velcro (hook & loop) tape so that a variety of badges could be affixed to the brassard to identify the wearer's duty role. A selection of these square duty badges is shown below.

The generic Air Force Blue brassard with hook-and-loop tape to affix various duty badges

Duty member

Health services personnel

Air movements

Military skills instructor

Chaplain of Christian faith

Chaplain of Jewish faith

Service Police

Corporals Graham Gleeson, Ivan Brezinscak, Tammy Dolby photographed together wearing the MSI brassard in Air Force Blue and All Seasons colours with matching uniforms. Changeovers take time and there is always an overlap to ensure that items are not discarded unnecessarily

RAAF UNIFORMS IN 2017

By 2017, the Air Force had a uniform wardrobe that clearly identified its wearer and was practical in environments ranging from Antarctica to the deserts of the Middle East. The dark blue uniform, so fondly remembered by many from World War II, Korea and Vietnam, had been re-established in the public eye as the uniform of its Air Force.

CHAPTER 8: RAAF BANDS

Bands have, for centuries, played an important role in military services — as an essential element of a parade and in presenting the image of the service to the general public. To carry out their role, RAAF bands have often worn unique ceremonial uniforms and this chapter will cover some of those.

Soon after the creation of the RAAF, a part-time RAAF Band was formed in February 1923 at RAAF Point Cook. As well as performing at military parades, the band's first major public engagement was at the opening of Parliament House in Canberra in 1927. A second part-time brass band was established at RAAF Richmond in 1932. Throughout World War II, the two bands performed parades and concerts all over the country, raising morale among service men and women, as well as among the public.

In 1952, the band based at Laverton became a full-time band, recruiting professional musicians who could play a wider range of music than could be performed by part-time bandsmen. It was renamed the RAAF Central Band and, in 1954, it supported the Royal Tour by providing music at regal functions in all states of Australia. The band at Richmond became full-time in 1969, later being renamed Air Command Band. The two bands were amalgamated in 2008 into one band named Air Force Band. In recent years, Air Force musicians have deployed to the Australian Defence Force's areas of operation, entertaining troops in East Timor, The Solomon Islands, Iraq and Afghanistan.

THE FIRST BAND UNIFORMS

1952 bandsman ceremonial dress

Until 1952, bandsmen in RAAF bands wore service dress uniform when performing. With the formation of the full-time RAAF Central Band in 1952, a new ceremonial uniform for the band was approved. It consisted of a dark blue button-to-the-chin tunic, semi-tight trousers with a light blue stripe and gold border, half Wellington boots and a shiny peaked service cap with gold braid on the peak.[67] A new ceremonial dress was approved for band officers, based on the full dress uniform worn in the later 1930s. Although shoulder boards with an eagle-and-crown badge were worn, gold rank insignia were worn on the sleeve for both officers and airmen.

In the late 1950s, a summer version of band ceremonial dress was introduced which replaced the dark blue tunic with a lighter white one. The ceremonial uniforms were designed for standing and marching, but the tunic was unsuitable for sitting in. In the early 1960s, a concert uniform was introduced consisting of the same trousers as the ceremonial uniform but with a short white jacket similar in style to the SNCOs mess dress jacket. The jacket was worn over a white shirt and black bow-tie. As concerts were always indoors, no hat was worn.

Summer ceremonial uniform in the 1960s

Central Band in summer ceremonial dress being inspected by HRH The Prince of Wales in 1983

Summer concert uniform from the 1960s

With the change to the all-season uniform in 1972, band members wore the all-seasons uniform for everyday wear but their ceremonial and concert uniforms were largely unchanged from the 1960s.

---

[67] Michael Armit, 'He's Our Mr Music', The Argus, Melbourne VCI 31 March 1956, viewed at trove.nla.gov.au/newspaper/article/71795027 on 31 March 2018

RAAF Central Band on parade at RAAF Point Cook in the 1980s. They are wearing their winter ceremonial dress

Band ceremonial dress cap from 1966 with gold braid on the peak

Band member ceremonial peak cap in 1987

Band officer ceremonial peak cap in 1987

Air Vice-Marshal (later, Air Chief Marshal Sir) Angus Houston in his Air Force Blue service dress speaking with Flight Lieutenant (later Squadron Leader) Steve Campbell-Wright at a social gathering in early 2001. Flight Lieutenant Campbell-Wright is wearing band officer's ceremonial dress which was based on the 1930s full dress uniform (Chapter 2)

With the changeover to Air Force Blue uniforms in 2000, the decision was made about 2002 to withdraw the ceremonial uniforms of band members in favour of the standard service dress uniform. Two items of the ceremonial uniform that continued to be worn were the drum major's blue sash and mace.

The Air Force Band on parade in their service dress uniform, under the direction of the drum major wearing his blue sash. The Director of Music, Squadron Leader Matthew Shelley is at left of the photograph

In 2014, a new concert dress uniform was introduced for band members. It was based on the Air Force Blue winter mess dress and was worn with a white shirt and black bow-tie.

The Air Force Band Wind Section in the new concert dress uniform in 2014

BADGES

The following badges were unique to band members:

Drum Major

The drum major is the leader of the marching band and controls pace and tempo. He is also responsible for drill and discipline and uses the mace to give commands. This important member of the band wore the drum major badge which came into use in the RAAF with the first band in the 1920s. The badge consisted of an inverted 4-bar chevron surmounted by a drum. On the white summer ceremonial uniform, the badge was in gold on a black background, but on winter uniform, it was white silk embroidered on dark blue.

Summer Uniform   *All Seasons Uniform*

Lyre Badge

The lyre badge used to identify musicians consisted of a bullion lyre on a black background. It was worn on both sleeves above the rank insignia on both summer and winter ceremonial dress. The lyre badge was withdrawn from use in 1989 in line with RAAF policy of personnel not wearing trade badges.

CHAPTER 9: CHAPLAINS

RAAF chaplains, being both religious clerics and members of the Air Force, held an unusual position and status which is reflected in the uniforms they have worn over nine decades. From 1926 when the first RAAF chaplain was appointed, they have worn officers' uniforms with distinctive insignia to identify their unique status.

FIRST CHAPLAINS' UNIFORM

In the late 1920s, the first chaplains wore the service dress of an officer with the following differences:

- The gilded cap badge worn by RAAF chaplains was similar to that used by chaplains of the Australian Army and the Australian Flying Corps. It consisted of a gilt Cross Pattee surmounted by a crown and was worn on the front of the service cap.
- A smaller version of the cap badge was worn on the sleeve above the rank lace in lieu of the eagle-and-crown badge.
- At least some of the chaplains wore the clerical collar instead of a tie with their uniform.[68]

Chaplain's hat badge and sleeve badge from the 1920s

THE 1930s

During the 1930s, three new items were introduced for chaplains. A lapel badge consisting of a Maltese Cross in black metal with wings of gilding metal with an 'RAAF' monogram, surrounded by a wreath superimposed over the centre of the cross. These were worn on both lapels when wearing winter or summer service dress tunics. When the field service cap was introduced in 1937, this badge was used unofficially as a badge on that cap and also on the khaki fur-felt hat. An oxidised version was introduced in the late 1930s.

The 1930s saw a new service dress cap badge come into use. The badge consisted of the gilt lapel badge surmounted by a crown in bullion embroidery on a black background. This badge remained in use until 1954 when the Tudor crown was changed to the St Edward's crown.

---

[68] Peter A. Davidson, Skypilot – A History of Chaplaincy in the RAAF 1926-1990, Department of Defence, Canberra ACT, 1990 p. 1-6

A chaplain's stole or scarf was also introduced in the 1930s to be worn during religious services. The stole was black and embroidered at both ends with a badge in red and gold of the same pattern as the cap badge.

The bullion embroidered badge from Chaplain's preaching stole (Protestant)

Chaplain's SD cap badge from 1930s to 1954

RAAF Chaplain's preaching stole (Protestant)

WORLD WAR II

It was clear that the huge wartime expansion of the RAAF would require additional Chaplains and, due to the location of bases in remote localities and overseas, those Chaplains would have to be full-time. Chaplains began working from temporary facilities, often tents or huts, in combat zones such as the Middle East or Papua New Guinea and the uniforms reflected the new environment. Khaki uniforms often replaced the dark blue service dress, but the badges and insignia remained the same.

Chaplains (Squadron Leaders) Bob Davies of Newcastle NSW (Church of England), John McNamara of Brunswick VIC (Roman Catholic) and Fred McKay of Queensland (Presbyterian) somewhere in Italy. All are wearing variations of khaki drab uniforms. Source: AWM Canberra

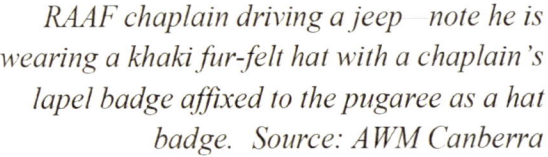
RAAF chaplain driving a jeep—note he is wearing a khaki fur-felt hat with a chaplain's lapel badge affixed to the pugaree as a hat badge. Source: AWM Canberra

POST-WAR

When Queen Elizabeth II ascended to the throne, she chose as her symbol the St Edward's Crown. As with other RAAF badges, the St Edward's Crown replaced the Tudor Crown on the Chaplain's cap badge and stole in 1954.

In 1955, the position of Principal Air Chaplain, for the most senior Chaplain in the RAAF, was established at Air Commodore rank, requiring the design of a Chaplain's version of the air officer's cap badge.[69]

Chaplain's cap badge from 1954 to 1972

Principal Chaplain's cap badge from 1955-72

Chaplain's stole (Protestant) with embroidered St Edward's Crown

---

[69] Peter A. Davidson, Skypilot – A History of Chaplaincy in the RAAF 1926-1990, Department of Defence, Canberra ACT, 1990 p. 10-7

Chaplain's cap fitted with 1954 to 1972 pattern chaplain's cap badge

Principal Air Chaplain's service dress cap from the 1950s

ALL SEASONS UNIFORM

With the introduction of the all-seasons uniform in 1972, chaplains ceased to wear their unique service dress cap badge in favour of the standard officer or air officer cap badge, as appropriate. A newly designed chaplain badge made its appearance at the same time and was initially worn on the shoulder boards in place of the eagle and crown badge. Later, this badge was worn as a lapel badge when sleeve rank was re-introduced.

Principal Air Chaplain's shoulder board from the 1970s

Chaplain's stoles (Anglican and/or Roman Catholic) in the purple, green, red and white colours for the Liturgical seasons

JEWISH CHAPLAINS

Following the appointment of the Air Force's first Jewish Chaplain in 2013, a new lapel badge for chaplains of that faith was introduced.

RAAF Chaplain Yossi Friedman (centre) blessed the Air Movements Training and Development Unit Governor-General's Banner in 2013 with (left to right) Chaplain Mark Willis, Principal Air Chaplain Kevin Russell and Principal Air Chaplain Peter O'Keefe

APPENDICES

1. Rank insignia – Commissioned Officers – Mid 1920s
2. Rank insignia – Warrant Officers, Non-Commissioned Officers and Airmen – 1921
3. Uniforms – Commissioned Officers & Warrant Officers Class I – 1921
4. Uniforms - Warrant Officers Class 2, Flight Sergeants and Sergeants – 1921
5. Uniforms – Corporals and Airmen – 1921
6. Rank insignia – Warrant Officers and Airmen – 1937
7. Uniforms – Commissioned Officers & Warrant Officers Class I – 1937
8. Uniforms – Airmen – 1937
9. Ranks and Special Insignia – 2015
10. Identified Manufacturers of Sterling Silver Flying Badges

Appendix 1: Rank Insignia – Commissioned Officers – Mid 1920s

Rank	Hat Badge & Peak	Service Dress (Winter)	Service Dress (Summer or Tropical)
Group Captain	Officer with one row of wattle leaves on peak of cap	4 rows of gold braid ½ inch wide around the sleeves of jacket below elbow, surmounted by a gilding metal eagle and crown	4 rows of gold braid ½ inch wide around the sleeves of jacket below elbow, surmounted by a gilding metal eagle and crown
Wing Commander	Officer with one row of wattle leaves on peak of cap	3 rows of gold braid ½ inch wide around the sleeves of jacket below elbow, surmounted by a gilding metal eagle and crown	3 rows of gold braid ½ inch wide around the sleeves of jacket below elbow, surmounted by a gilding metal eagle and crown
Squadron Leader	Officer with one row of wattle leaves on peak of cap	2 rows of gold braid ½ inch wide with one row of gold braid ¼ inch wide around the sleeves of jacket below elbow, surmounted by a gilding metal eagle and crown	2 rows of gold braid ½ inch wide with one row of gold braid ¼ inch wide around the sleeves of jacket below elbow, surmounted by a gilding metal eagle and crown
Flight Lieutenant	Officer	2 rows of gold braid ½ inch wide around the sleeves of jacket below elbow, surmounted by a gilding metal eagle and crown	2 rows of gold braid ½ inch wide around the sleeves of jacket below elbow, surmounted by a gilding metal eagle and crown
Flying Officer	Officer	1 row of gold braid ½ inch wide around the sleeves of jacket below elbow, surmounted by a gilding metal eagle and crown	1 row of gold braid ½ inch wide around the sleeves of jacket below elbow, surmounted by a gilding metal eagle and crown
Pilot Officer	Officer	1 row of gold braid ¼ inch wide around the sleeves of jacket below elbow, surmounted by a gilding metal eagle and crown	1 row of gold braid ¼ inch wide around the sleeves of jacket below elbow, surmounted by a gilding metal eagle and crown
Air Cadet	Officer	Gilding metal eagle and crown, small	Gilding metal eagle and crown, small

The information in this schedule is collated from an undated document entitled "Standing Orders (Provisional) for the Clothing of the Royal Australian Air Force". As it includes reference to the Citizen Air Force it is assumed that it was prepared in the mid-1920s. However, it remains the earliest recorded document dealing with uniforms.

Appendix 2: Rank Insignia – Warrant Officers, Non Commissioned Officers And Airmen – 1921

Rank	Hat Badge	Service Dress (Winter)	Service Dress (Summer and Tropical)
Warrant Officer Class I	Officer	Royal Arms embroidered in blue silk worn below elbows on both arms	Royal Arms in gilding metal worn below elbow on both arms
Warrant Officer Class II	Airman	Blue silk embroidered crown worn below elbow on both arms	Gilding metal crown worn below elbow on both arms
Flight Sergeant	Airman	3 blue chevrons surmounted by a blue silk 4 bladed propeller with star in centre worn above the elbow on both arms	3 drab chevrons surmounted by a drab silk 4 bladed propeller with star in centre worn above the elbow on both arms
Sergeant	Airman	3 blue chevrons surmounted by a blue silk 4 bladed propeller in centre worn above the elbow on both arms	3 drab chevrons surmounted by a drab 4 bladed propeller in centre worn above the elbow on both arms
Corporal	Airman	2 blue chevrons worn above elbow on both arms	2 drab chevrons worn above elbow on both arms
Leading Aircraftman	Airman	2-bladed airscrew in blue silk worn above the elbows on both arms	2-bladed airscrew in drab worn above the elbows on both arms
Aircraftman 1 Aircraftman 2 Boy	Airman	None	None

Appendix 3: Uniforms – Commissioned Officers & Warrant Officers Class I – 1921

Item	Service Dress (Winter)	Service Dress (Summer or Tropical)	Mess Dress
Cap	Of blue cloth. Peak of same material as cap, for officers of Wing Commander and below. A band of black mohair braid is worn around the cap.	As winter but with a cap cover of the same material as Jacket in this section	As for Service Dress (Winter).
Jacket	Same pattern as Army Service Dress, four buttoned front, but without shoulder straps. Belt of same material as Jacket and fastened with a 2 prong gilt buckle. Buttons to be gilt, plain edged and bearing a device of eagle surmounted by a crown. Cloth of approved blue colour in 18 ounce weight.	Same pattern as Army Service Dress but with detachable shoulder straps. Material to be drab drill of approved shade. Belt of same material as Jacket fastened with a 2-prong buckle. Buttons to be gilt or gilding metal, plain edged and bearing a device of eagle surmounted by a crown. Badges of rank will be worn on detachable shoulder straps. *Officers who so desire may substitute Gabardine for drab drill in this section providing it is of approved shade.*	Same pattern as Royal Air Force (Eton style) in the approved blue colour with facings of light blue. Buttons to be gilt or gilding metal, plain edged and bearing a device of eagle surmounted by a crown. Badges of rank will be worn on shoulder straps and to include gilt eagle and crown.
Trousers	Of blue cloth. Ordinary service pattern. With or without 'turn-ups'. Lap seams approximately 1/8 inch in width.	Drab drill of approved shade with permanent turned up bottoms.	Same pattern as Royal Air Force (Eton style) in the approved blue colour with leg stripes of light blue.
Breeches	Army Mounted Service pattern in blue. Strappings to be of same material as breeches.	Army Mounted Service pattern in drab colour –may be either drill or Bedford Cord. Strappings to be of same material as breeches.	

Appendix 3: Uniforms – Commissioned Officers & Warrant Officers Class I – 1921 (continued)

Item	Service Dress (Winter)	Service Dress (Summer or Tropical)	Mess Dress
Puttees		Army pattern but drab in colour	
Collar	White. May be stand and fall, wing polo or soft. Gold pin may be worn with soft collar.	Drab in colour of same material as shirt. Gold pin may be worn with soft collar.	
Shirt	White, soft or stiff front.	Drab in colour, Linen, flannel or suitable material.	White dress (wing pointed)
Tie	Black silk (knitted) or poplin. End of tie to be approximately 2 inches in width.		Black bow tie.
Helmet		Wolseley Pattern fitted with puggaree	
Hat		Fur Felt, Australian pattern with puggaree. Hat will be worn turned down all around	
Puggaree		Drab colour. 7 folds, the fourth being dark blue	
Shoes	Black leather, plain toecap		
Socks	Black.		
Gloves	Brown leather or suede		
Boots, Ankle	Black leather, plain with 9 pairs of eyelet holes.		
Boots, Field	For ranks Squadron Leader and above. Black leather, plain toecap. Straight back and laced instep with 9 pairs of eyelet holes. Separate leg strap running through a small leather loop at the back of boot.		
Walking Stick	Yellow cane with crook.		

Appendix 3: Uniforms – Commissioned Officers & Warrant Officers Class I – 1921   (continued)

Item	Service Dress (Winter)	Service Dress (Summer or Tropical)	Mess Dress
Coat, Rain	Any recognised military pattern but blue in colour without shoulder straps and to be fitted with a belt of the same material and with a suitable buckle.	When not on parade, coat of khaki or drab colour may be worn with summer dress.	
Greatcoat	Blue, Royal Air Force pattern. Buttons to be gilt or gilding metal, plain edged and bearing a device of eagle surmounted by a crown. Belt to be of same material (described as 30 ounce cloth of the approved blue colour) as greatcoat, fitted with a sliding buckle of the approved design. Detachable shoulder straps of same material as greatcoat.		
Coat (British Warm)	Blue, Army pattern but with a belt and sliding buckle of Gilt metal. Buttons as greatcoat. Detachable shoulder straps as for greatcoat. Material to be 30 ounce cloth as for Greatcoat. Officers in possession of Khaki British Warm may have same dyed to correct shade.		

Service Dress Caps: Officers usually had two caps, one used as a working dress cap without embellishments, and a parade/ceremonial cap, which included all embellishments.

Mess Dress Waistcoat: Same pattern as Royal Air Force (Eton style) in the approved blue colour. Buttons (4) to be gilt or gilding metal, plain edged and bearing a device of eagle surmounted by a crown.

Tropical Mess Dress: Mess Jacket, trousers (cut as overalls) made in white drill and vest and cap cover made of white Marcella material.

Mess Dress Flying Badges: Officers who are qualified to wear the flying badge or the observer's badge, will wear a miniature of the Full Dress badge on the left lapel.

Appendix 4: Uniforms - Warrant Officers Class 2, Flight Sergeants and Sergeants – 1921

Item	Service Dress (Winter)	Service Dress (Summer or Tropical)
Cap	Field Service (similar to RAF. Later called: Cap, Service). Crown of blue cloth. Peak to be cloth covered. Black braid band. Black glazed leather chinstrap fastened by 2 small gilt buttons on side and top of peak. Material to be same as for jacket.	As winter but with a cap cover of the same material as Jacket in this section
Jacket	Blue cloth with turn down collar, 2 top patch pockets, 2 side slit and flap pockets. Cloth belt with 2-pronged gilding metal buckle. Plain edged gilding metal buttons bearing the device of an eagle surmounted by a crown. Material to be 18 oz. cloth of approved blue colour	As per Winter uniform but in drab drill instead of blue cloth.
Collar	White, soft, linen. Gold pin may be worn with soft collar.	Drab, linen or flannel.
Shirt	White, soft.	
Tie	Black cotton or black silk knitted	Black cotton or black silk knitted
Trousers	Blue with permanent turned up bottoms.	Drab drill with permanent turned up bottoms.
Breeches	Army Mounted Service pattern in blue. Strappings to be of same material as breeches.	As per Winter uniform but in drab drill instead of blue cloth.
Great Coat	Blue, Royal Air Force pattern. Buttons to be gilt or gilding metal, plain edged and bearing a device of eagle surmounted by a crown. Belt to be of same material (described as 30 ounce cloth of the approved blue colour) as greatcoat, fitted with a sliding buckle of the approved design. Detachable shoulder straps of same material as greatcoat.	As per Winter uniform.
Leggings	Black leather, plain.	Black leather, plain.
Boots	(Ankle) Black leather, plain	Black leather, plain
Hat		Fur felt Australian pattern fitted with puggaree. Brim to be worn turned down all round.

Appendix 5: Uniforms – Corporals and Airmen – 1921

Item	Service Dress (Winter)	Service Dress (Summer or Tropical)
Cap	Field Service (similar to RAF. Later called Cap, Service). Crown of blue cloth. Peak to be cloth covered. Black braid band. Black glazed leather chinstrap fastened by 2 small gilt buttons on side and top of peak. Material to be same as for jacket.	As winter but with a cap cover of the same material as Jacket in this section.
Jacket	Blue. Stand and Fall collar. 2 top patch pockets, 2 side slit and flap pockets. Cloth belt with 2 pronged gilding metal buckle. Plain edged gilding metal buttons bearing the device of an eagle surmounted by a crown. No shoulder straps. Material to be 18 oz. cloth of approved blue colour.	As per Winter uniform but in drab drill.
Trousers	Blue with permanent turned up bottoms.	Drab drill with permanent turned up bottoms.
Breeches	Army Mounted Service pattern but blue in colour. Strappings of same material as breeches. Material to be 24 oz. Bedford cord of approved blue shade.	Army Mounted Service pattern in drab drill of same material as breeches.
Greatcoat	Similar to British Army Mounted Service pattern (other ranks) but blue in colour. Buttons to be of gilt or gilding metal plain edged and bearing the device of an eagle surmounted by a crown. No shoulder straps. Material to be 30 oz. cloth of approved blue shade.	As per Winter uniform.
Leggings	Black leather, plain	Black leather, plain
Boots, Ankle	Black leather, plain	Black leather, plain
Hat		Fur felt Australian pattern fitted with puggaree. Brim to be worn turned down all round.

Appendix 6: Rank Insignia – Warrant Officers and Airmen – 1937

Rank	Hat Badge	Service Dress (Winter)	Service Dress (Summer and Tropical)
Warrant Officer	Warrant Officer	Royal Arms in gilding metal worn below elbow 6½ inches from bottom of sleeve to centre of badge, on both sleeves.	Royal Arms in gilding metal worn below elbow 6½ inches from bottom of sleeve to centre of badge, on both sleeves. When jacket is not worn the badge is to be worn on a strap on the right wrist.
Warrant Officer Class II	Airmen	Brass crown worn below elbow, on both sleeves.	Oxidised crown worn below elbow, on both sleeves.
Flight Sergeant	Airmen	Chevrons, 3-bar worsted surmounted by a metal (gilt) crown worn above the elbow lowest point of chevrons 9½ inches from top of sleeve. Crown ¾ inch above chevrons. Worn on both sleeves.	Chevrons, 3-bar worsted surmounted by a metal (oxidised) crown worn above the elbow lowest point of chevrons 9½ inches from top of sleeve. Crown ¾ inch above chevrons. Worn on both sleeves.
Sergeant	Airmen	Chevrons, 3-bar worsted worn above the elbow lowest point of chevrons 9½ inches from top of sleeve.	Chevrons, 3-bar worsted worn above the elbow lowest point of chevrons 9½ inches from top of sleeve.
Corporal	Airmen	Chevrons, 2-bar worsted worn above the elbow lowest point of chevrons 9½ inches from top of sleeve.	Chevrons, 2-bar worsted worn above the elbow lowest point of chevrons 9½ inches from top of sleeve.
Leading Aircraftman	Airmen	Airscrew, metal (gilt).	Airscrew, metal (oxidised).
Aircraftman 1, 2 & Boy	Airmen	None	None

Warrant Officers, Class II (also called Sergeant-Major, 2nd Class were to remain on the rank structure until 15 February 1940 (Statutory Rule dated 24 July 1941) when the ranks of Sergeant-Major, 1st Class and Sergeant-Major, 2nd Class were merged into that of Warrant Officer with seniority to date from the date first promoted to Sergeant-Major, 2nd Class.

The colour of worsted badges shall be that of the garment on which worn.

Appendix 7: Uniforms – Commissioned Officers & Warrant Officers Class I – 1937

Item	No. 1 Full Dress (Winter)	No. 2 Full Dress (Tropical)	No. 3 Mess Dress
Headwear	Headdress, RAF Pattern. Officers of Air Rank to wear a special plume in lieu of "flash".	Headdress, RAF Pattern [1]. Officers of Air Rank to wear a special plume in lieu of "flash".	Cap, Service Dress, officer's with two buttons, chin strap, black mohair band.
Tunic	Tunic, blue, full dress. These tunics included special collars and shoulders straps for each officer grouping. (See below)	Tunic, white, Full dress	Jacket, Mess. Waistcoat, mess, white. Shirt, evening, white, stiff, plain front, one or two stud-holes, plain gold studs and links. Black bow tie.
Overalls	Blue	White, or blue as ordered	Blue
Boots	Boots, Half Wellington	Boots, Half Wellington	Boots, Half Wellington
Gloves	Gloves, white	Gloves, white	Gloves, white
Shirt	Shirt, white, stiff cuffs	Shirt, white	As above
Sword	Sword. RAF Pattern (Not carried by WOs)	Sword. RAF Pattern	

Special Collars and Shoulder Straps for the No 1 Full Dress Tunic (Winter) of each Officer Grouping.

Air Officers	GPCAPT, WGCDR SQNLDR	Flight Lieutenant and below
Collar – Device of oak leaves and acorns in gold embroidery running all round the collar.	Collar – Device as per officers of air rank, but terminating 3½ inches from front of each side of collar.	Collar – Device of five oak leaves in gold embroidery in front of each side of collar.
Shoulder device – Eagle and crown in gold embroidery. Laurel wreath embroidered in gold round button at top of shoulder strap.	Shoulder device – Eagle and crown as per officers of air rank but without laurel wreath.	Shoulder device – As per Group Captain etc.

1 Air Board minute of 1935 mentions a white helmet to be worn with this version of dress but, as it was a private purchase item for officers, there is a distinct possibility that neither the uniform nor the headwear was ever made.

Appendix 8: Uniforms – Airmen – 1937

Item	No. 5 Service Dress	No. 5A Field Service Dress	No. 6 Service Dress (Summer)	No. 6A Field Service Dress (Summer)	No. 6B Tropical Dress
Headwear	Cap, service dress, airman's with two buttons, chinstrap and black mohair band.	Cap, field service, airman's	Hat, fur felt, or cap, service dress, airman's.	Cap, field service, airman's	Helmets, drab, complete with cover, waterproof; puggaree with flash, and strap, chin or hat, fur felt, or cap, service dress or field service as ordered.
Jacket	Jacket, blue, airman's	Jacket, blue, airman's	Jacket, drab, airman's	Jacket, drab, airman's	Jacket, drab, airman's
Trousers (Shorts)	Trousers, blue (not turned up), airman's	Trousers, blue (turn-ups optional), airman's	Trousers, drab (not turned up), airman's	Trousers, drab (turn-ups optional), airman's	Shorts, drab
Shirt	Blue-grey, airman's	Blue-grey, airman's	Bush, airman's	Bush, airman's	Bush, airman's
Collar	Blue-grey, soft	Blue-grey, soft			
Tie	Black, airman's pattern	Black, airman's pattern	Black, airman's pattern	Black, airman's pattern	
Boots	Ankle, black, airman's, leather or rubber soles with laces		Ankle, black, leather or rubber soles with laces		
Socks	Socks, Black	Socks, Black	Socks, Black	Socks, Black	Stockings, drab
Shoes	Black, airman's, leather or rubber soles with laces	Black, airman's, leather or rubber soles with laces	Black, airman's, leather or rubber soles with laces	Black, airman's, leather or rubber soles with laces	Black, airman's, leather or rubber soles with laces

Appendix 9: RANKS AND SPECIAL INSIGNIA – 2015 **Source: Department of Defence**

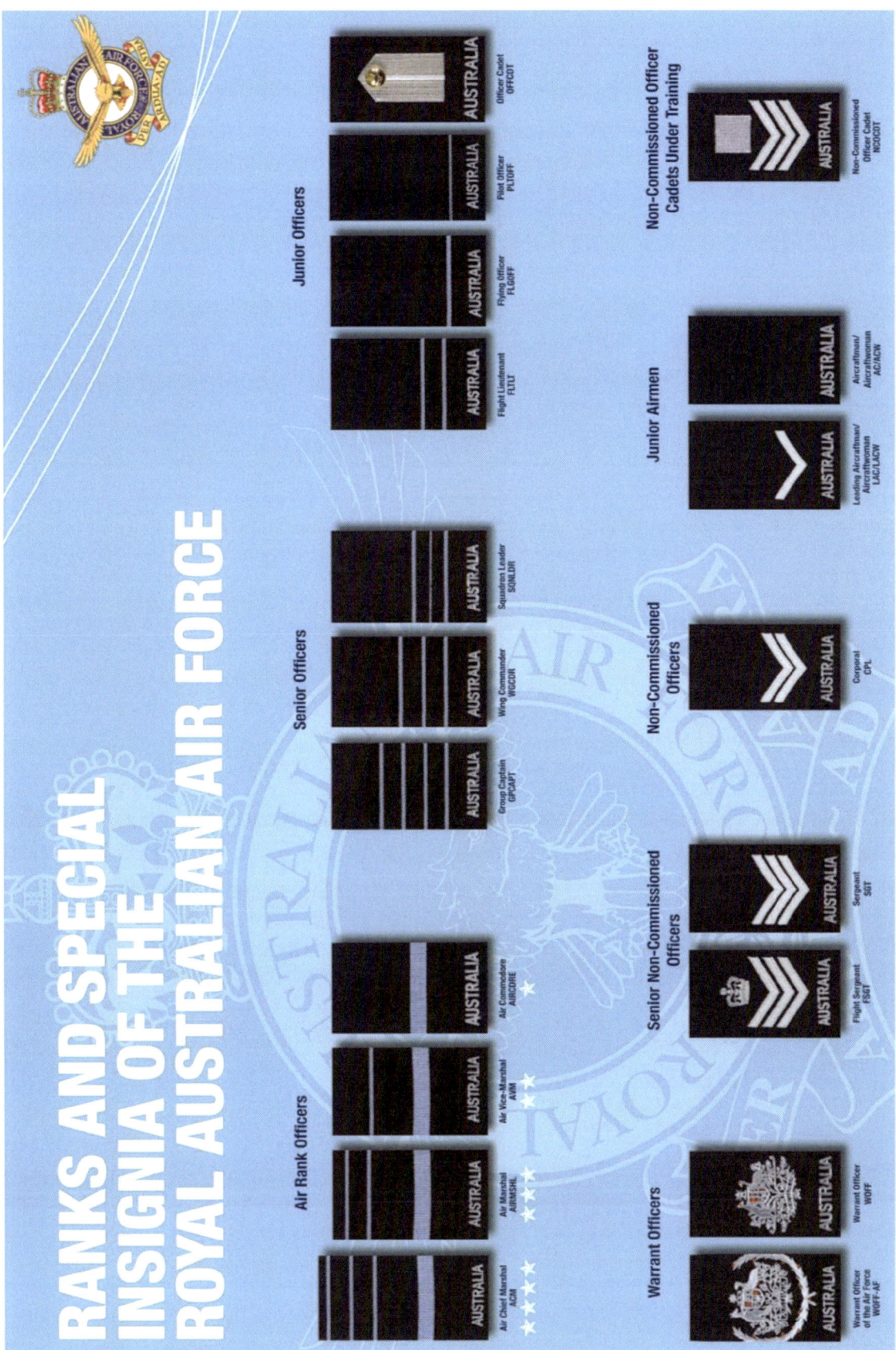

Appendix 10: Identified Manufacturers of Sterling Silver Flying Badges

This table has been derived from information published privately in *RAAF STERLING SILVER BREVETS* by Ian Jenkins, Dennis Graham & John Macdonald ©2009, and revised by Ian Jenkins & Dennis Graham ©2022. These publications record examples which are definitely known in collections and are not meant to provide a definitive list of all combinations which do exist.

All RAAF crowned metal flying brevets are made of sterling silver and are therefore required to have a sterling silver mark, which is always found on the reverse. The marks found are STERLING, STG or STG SIL, with or without spaces and full stops. All the earlier badges also carry makers' marks, sometimes with dates, but manufacturing details and dates ceased to be added on later badges. The latest date to be included is 1980 on LEGA manufactured Pilot and Flight Steward wings. An Airborne Electronics (AE) wing with clutch pins is also known without a sterling mark or manufacturer/date details and this is possibly a reproduction.

The basic design of the sterling flying badges has remained constant since their introduction in 1949, with a change from Tudor crown to St Edward's crown from 1954 and a later move from brooch pins to clutch pins. None of the clutch pin badges have a manufacturing date on the reverse.

Tudor Crown

Five flying badges were manufactured with a Tudor crown: Pilot (RAAF), Navigator (N), Signaller (S), Engineer (E) and Gunner (G). They were only made with brooch fittings and all are found with makers' marks for 'Stokes & Sons, 1949' and 'K.G. Luke, Melbourne, 1952'.

Parachute Jump Instructors

An uncrowned Parachute Jump Instructor badge was introduced in 1951 with a parachute symbol in the wreath. This badge was made in silver plated or silver washed, bronze or brass, rather than sterling silver. It had a brooch pin, no crown and no maker's mark. A crowned version in sterling silver was introduced in 1955 as part of the programme to replace the Tudor crown with the St Edward's crown on all flying badges. Responsibility for parachute jump instruction passed to the Army in 1972 and so it is no longer an RAAF trade. The design of the Army parachute instructor badge is the same as the RAAF design, but the badges manufactured for the Australian Army are not made in silver, but in bronze, silver (white metal) and gold finishes to reflect specific qualifications. RAAF sterling silver Parachute Jump Instructor badges are listed in the table of St Edward's crown flying badges, below.

Airborne Electronics (AE, 1967)

Originally for Airborne Electronics Officers and later changed to Airborne Electronics Analyst.

Southern Cross Brevet 1998

The Southern Cross brevet with an image of the 5 stars in the Southern Cross constellation, was originally introduced in 1998 for commissioned aircrew: Navigator, Airborne Electronics Officer and Fighter Controller. The name was later changed to Air Combat Officer and subsequently to Mission Aircrew.

St Edward's Crown

Qualification	Symbol	Fitting	Maker/Manufacturer's Mark
Pilot	**RAAF**	Brooch	Lega 1978, Lega 1980, Lega Melb, K G Luke, Lustre Melb 1962, Perfection Sydney, Also with no maker's mark or date.
		Clutch	No maker's mark or date. Also known as a sealed sample by Dylmain Pty Ltd
Navigator	**N**	Brooch	Lega 1979, Millers, Stokes & Sons 1955, Swann & Hudson
		Clutch	Dylmain P/L, Nichol. Also with no maker's mark/date
Signaller	**S**	Brooch	K G Luke, Stokes & Sons 1955, Swann & Hudson Frankston
Gunner	**G**	Brooch	Brim, Brim Melb, Lega 1979, K G Luke, K G Luke Melbourne, Millers, Stokes & Sons 1955.
		Clutch	No maker's mark or date
Engineer	**E**	Brooch	Brim Melb, K G Luke Melbourne, Stokes Melb, Swann & Hudson.
		Clutch	No maker's mark or date.
Parachute Jump Instructor*	(parachute)	Brooch	K G Luke
		Clutch	K G Luke. Also no maker's mark or date.
Loadmaster 1965	**L**	Brooch	Stokes (A'Asia), Swann & Hudson Frankston
		Clutch	No maker's mark or date.
Crewman 1966	**C**	Brooch	Stokes, Stokes (A'Asia)
Air Electronics 1967	**AE**	Brooch	K G Luke, Stokes (A'Asia). Also with no maker's mark or date.
		Clutch	No maker's mark or date. Also with no maker's mark or date and known as sealed sample by Dylmain Pty Ltd
Flight Steward 1975	**FS**	Brooch	Lega 1980, K G Luke, Swann & Hudson. Also with no maker's mark or date.
		Clutch	No maker's mark or date
Crewman Technical 1983	**CT**	Brooch	Lega Melbourne
Technical Crewman 1983	**TC**	Brooch	Swann & Hudson Melb. Manufactured in error
Fighter Controller 1996	**FC**	Clutch	No maker's mark or date
Southern Cross 1998	**Stars**	Clutch	No maker's mark or date
Flight Technician 1999	**FT**	Clutch	Nichol. Also with no maker's mark or date.
Crew Attendant 2002	**CA**	Clutch	No maker's mark or date.

BIBLIOGRAPHY

Books

Coulthard-Clark, C.D., *The Third Brother – The Royal Australian Air Force 1921-39*, Allen and Unwin, North Sydney NSW, 1991

Davidson, Peter A, *Skypilot – A History of Chaplaincy in the RAAF 1926-1990*, Department of Defence, Canberra ACT, 1990

Department of Air, *The Golden Years; Royal Australian Air Force 1921–1971*, Australian Government Publishing Service, Canberra ACT, 1971

Friedlander, Walter J., *The Golden Wand of Medicine: A History of the Caduceus Symbol in Medicine*, Greenwood Press, Santa Barbara CA, 1992

Halstead, Gay, *The Story of the RAAF Nursing Service*, Nungurner Press, Metung VIC, 1994

Hering, P.G., Squadron Leader RAF, *Customs and Traditions of the Royal Air Force*, Gale and Polden Ltd, Aldershot, Hampshire UK, 1961

Hobart, Malcolm C,. Badges and Uniforms of the Royal Air Force, Pen and Sword Books Ltd, Barnsley, South Yorkshire UK, 2000

Philpott, Ian, *The Royal Air force: An Encyclopedia of the Inter-War Years 1930–1939*, Pen and Sword Publishing, Barnsley, South Yorkshire UK, 2006

O'Connor, John, *Australian Airborne: The History and Insignia of Australian Military Parachuting*, John O'Connor, Kingsgrove, NSW, 2006

Royal Australian Air Force, *Customs of the Royal Australian Air Force—A Concise History, Fourth Edition,* Royal Australian Air Force, Canberra ACT, 1996

Royal Australian Air Force, *RAAF Manual of Dress AAP 5135.003,* Royal Australian Air Force, Canberra ACT, 30 July 2015

Royal Australian Air Force, *Units of the Royal Australian Air Force,* AGPS, Canberra ACT, 1995

Stephens, Dr Alan, *The Australian Centenary History of Defence, Vol II, The Royal Australian Air Force*, Oxford University Press, Melbourne VIC, 2001

Stephens, Alan, *Going Solo, The Royal Australian Air Force 1946–1971,* Australian Government Publishing Service, Canberra ACT, 1995

Stephens, Alan and Jeff Isaacs, *High Flyers–Leaders of the Royal Australian Air Force*, AGPS, Canberra ACT, 1996

Walker, Allan S., *Medical Services of the RAN and RAAF,* Australian War Memorial, Canberra ACT, 1961

Williams, Sir Richard, Air Marshal (Retired), *These are Facts – The Autobiography of Air Marshal Sir Richard Williams, KBE, CB, DSO*, Australian War Memorial and Australian Government Publishing Service, Canberra ACT, 1977

Archives

Air Board Agenda records held by Directorate of History–Air Force, Canberra ACT Air Board Orders archives held by Directorate of History–Air Force, Canberra ACT

Air Force Headquarters Files

OCAF 2000/39507 parts 4–7, RAAF Uniform Policy

OCAF 2000/17839 parts 5 and 7, Uniform Design and Development

OCAF 2001/28570 part 2, Approval for New Uniform Items OCAF 2005/1015670 parts 3–6, Unit Caps,

Logos and Tee shirts OCAF/2007/1057003 part 2, Letters to the Public

OCAF 2000/13229 parts 4 and 5, Minutes of the Clothing Steering Committee

OCAF 2003/46756 part 1, Commanders Net Notices OCAF 2002/7798 parts 1–3, Air Force Badges and Brevets OCAF 2007/1098876 part 1, Dress Waivers

OCAF 2005/120560 part 3, Briefs for DCAF and CAF

OCAF 2008/1007680 part 1, Introduction of leather jacket

Journals, Magazines and Papers

Armit, Michael, 'He's Our Mr Music', *The Argus*, Melbourne VCI 31 March 1956, viewed at *trove.nla.gov.au/newspaper/article/71795027* on 31 March 2018

Baldwin, Ed, Squadron Leader, 'Historical Look at a Topical Problem', *RAAF News*, Canberra ACT, February 1983

Jenkins, Ian D., 'The P.N.B.W. Aircrew Scheme in the RAAF', unpublished paper, Canberra ACT, 2008

Simpson, Douglas, 'Timeline of Australian Defence Force Brevets 1914–2014,' self-published, Perth WA, 2014

Department of Defence, *Air Force News* and its predecessor, *RAAF News*, Canberra ACT, various years' editions.

Websites and Electronic Media

Anne Heywood, Women's Auxiliary Australian Air force (WAAAF) 1941-1947, The Australian Women's Register website, *www.womenaustralia.info/biogs/AWE0400b.htm* 3 December 2002

Australian War Memorial website, *www.awm.gov.au*

Encyclopedia Britannica website, *www.britannica.com/topic/caduceus*

Imperial War Museum website, *www.iwm.org.uk*

Royal Australian Air Force Association, WRAAF Branch website, History of the Women's Royal Australian Air Force 1951–1977, 2009, *http://www.wraaf.org.au/History.html viewed 1 March 2018*

RAAF Museum website, *www.airforce.gov.au/RAAFMuseum*

The Wadham's family website, WAAF Service Dress Uniform Buyer's Guide, dated 16 November 2015, *www.wadhamsfamilyhistory.co.uk/FortiesWAAFuniform.htm,* viewed 10 March 2018

JOHN MACDONALD

John Macdonald was introduced to the collecting of militaria by his father while still a young boy. After over thirty years, this habit developed into a passion. Some time ago, ever eager to expand on his collection, John sought a new challenge and decided to put together a display of World War II Commonwealth Air Force cap and aircrew badges. This seemingly simple task was to take John down the path which eventually led to the writing of *Shades of Blue*.

In all his previous experience, John had found that there existed a diverse library of reference material detailing a wide range of militaria in terms of uniforms and badges. However, he was surprised to find that there were very few published sources dealing with air force uniforms in general and those of the Royal Australian Air Force in particular. As a result, John and fellow collectors began to establish their own reference collections and knowledge on the subject.

It soon became clear that there was a need to consolidate their joint knowledge of RAAF uniforms into a format that could be used by collectors in both Australia and internationally. John and his colleagues submitted a proposal to the then Chief of Air Force, Air Marshal Mark Binskin, to write such a reference book. To the surprise and delight of the group the proposal was accepted. What was not envisaged was the length of time that the project would take to bring to fruition. However, after many long hours and much hard work, the manuscript was completed in early 2016 and was entered into that year's Air Force Heritage Awards. In recognition of the heritage value of the book to the RAAF it was awarded second prize in the Literary Section.

The enduring legacy of this book is not just in the information in these pages, but also that the RAAF now has a reference work which can be expanded on as our uniforms continue to evolve and as research reveals hitherto forgotten secrets of our past. For that gift, John Macdonald has the RAAF's eternal gratitude.

POSTSCRIPT

John mentions the long gestation period for the publication of this book which unfortunately has been extended long beyond his expectations. A fully edited manuscript was ready for publication in 2016 when it was submitted for the Air Force Heritage Awards and, having agreed to publish the book, John began to work with the RAAF to revise the manuscript. This collaboration improved the content in a number of areas, but also removed a number of elements, particularly affecting badges and insignia, in order to reduce the overall size of the publication. John battled through a prolonged illness over this period and to his delight, on 20 July 2018, the RAAF presented him with a printed and bound copy of Shades of Blue. John died four days later on 24 July 2018. It was a great disappointment to John's son Hugh, and the many enthusiasts who were eagerly awaiting its publication, that the RAAF subsequently decided not to proceed with the publication and it was 2021 before Hugh felt able to consider proceeding with his father's book. The project was further delayed by restrictions imposed by COVID-19 and the need to completely review and reset the RAAF manuscript. However, as a result, it has been possible to reintroduce much of the material which had been removed and I must pay my own tribute to the assistance I have received from Lindsay Cheal and Dennis Graham. I am delighted that the book and all John's research is now able to be made available to everyone with an interest in the uniforms and the insignia of the RAAF.

John Murdoch (Editor)   *June 2023*

www.ingramcontent.com/pod-product-compliance
Ingram Content Group UK Ltd
Pitfield, Milton Keynes, MK11 3LW, UK
UKRC040338240426
12049UKWH00017B/168